T0302074

CAMBRIDGE TRACTS IN MATHEMATICS

General Editors

J. BERTOIN, B. BOLLOBÁS, W. FULTON, B. KRA,
I. MOERDIJK, C. PRAEGER, P. SARNAK, B. SIMON, B. TOTARO

224   Attractors of Hamiltonian Nonlinear Partial Differential Equations

CAMBRIDGE TRACTS IN MATHEMATICS

GENERAL EDITORS

J. BERTOIN, B. BOLLOBÀS, W. FULTON, B. KRA, I. MOERDIJK, C. PRAEGER, P. SARNAK, B. SIMON, B. TOTARO

A complete list of books in the series can be found at www.cambridge.org/mathematics. Recent titles include the following:

190. Jordan Structures in Geometry and Analysis. By C.-H. CHU
191. Malliavin Calculus for Lévy Processes and Infinite-Dimensional Brownian Motion. By H. OSSWALD
192. Normal Approximations with Malliavin Calculus. By I. NOURDIN and G. PECCATI
193. Distribution Modulo One and Diophantine Approximation. By Y. BUGEAUD
194. Mathematics of Two-Dimensional Turbulence. By S. KUKSIN and A. SHIRIKYAN
195. A Universal Construction for Groups Acting Freely on Real Trees. By I. CHISWELL and T. MÜLLER
196. The Theory of Hardy's Z-Function. By A. IVIĆ
197. Induced Representations of Locally Compact Groups. By E. KANIUTH and K. F. TAYLOR
198. Topics in Critical Point Theory. By K. PERERA and M. SCHECHTER
199. Combinatorics of Minuscule Representations. By R. M. GREEN
200. Singularities of the Minimal Model Program. By J. KOLLÁR
201. Coherence in Three-Dimensional Category Theory. By N. GURSKI
202. Canonical Ramsey Theory on Polish Spaces. By V. KANOVEI, M. SABOK, and J. ZAPLETAL
203. A Primer on the Dirichlet Space. By O. EL-FALLAH, K. KELLAY, J. MASHREGHI, and T. RANSFORD
204. Group Cohomology and Algebraic Cycles. By B. TOTARO
205. Ridge Functions. By A. PINKUS
206. Probability on Real Lie Algebras. By U. FRANZ and N. PRIVAULT
207. Auxiliary Polynomials in Number Theory. By D. MASSER
208. Representations of Elementary Abelian p-Groups and Vector Bundles. By D. J. BENSON
209. Non-homogeneous Random Walks. By M. MENSHIKOV, S. POPOV, and A. WADE
210. Fourier Integrals in Classical Analysis (Second Edition). By C. D. SOGGE
211. Eigenvalues, Multiplicities and Graphs. By C. R. JOHNSON and C. M. SAIAGO
212. Applications of Diophantine Approximation to Integral Points and Transcendence. By P. CORVAJA and U. ZANNIER
213. Variations on a Theme of Borel. By S. WEINBERGER
214. The Mathieu Groups. By A. A. IVANOV
215. Slenderness I: Foundations. By R. DIMITRIC
216. Justification Logic. By S. ARTEMOV and M. FITTING
217. Defocusing Nonlinear Schrödinger Equations. By B. DODSON
218. The Random Matrix Theory of the Classical Compact Groups. By E. S. MECKES
219. Operator Analysis. By J. AGLER, J. E. MCCARTHY, and N. J. YOUNG
220. Lectures on Contact 3-Manifolds, Holomorphic Curves and Intersection Theory. By C. WENDL
221. Matrix Positivity. By C. R. JOHNSON, R. L. SMITH, and M. J. TSATSOMEROS
222. Assouad Dimension and Fractal Geometry. By J. M. FRASER
223. Coarse Geometry of Topological Groups. By C. ROSENDAL
224. Attractors of Hamiltonian Nonlinear Partial Differential Equations. By A. KOMECH and E. KOPYLOVA

Attractors of Hamiltonian Nonlinear Partial Differential Equations

ALEXANDER KOMECH
Universität Wien

ELENA KOPYLOVA
Universität Wien

CAMBRIDGE
UNIVERSITY PRESS

CAMBRIDGE
UNIVERSITY PRESS

University Printing House, Cambridge CB2 8BS, United Kingdom

One Liberty Plaza, 20th Floor, New York, NY 10006, USA

477 Williamstown Road, Port Melbourne, VIC 3207, Australia

314–321, 3rd Floor, Plot 3, Splendor Forum, Jasola District Centre, New Delhi – 110025, India

103 Penang Road, #05–06/07, Visioncrest Commercial, Singapore 238467

Cambridge University Press is part of the University of Cambridge.

It furthers the University's mission by disseminating knowledge in the pursuit of education, learning, and research at the highest international levels of excellence.

www.cambridge.org
Information on this title: www.cambridge.org/9781316516911
DOI: 10.1017/9781009025454

First published 2022

A catalogue record for this publication is available from the British Library.

ISBN 978-1-316-51691-1 Hardback

In memory of Mark Vishik

Contents

Preface   *page* ix

Introduction   1

1   Global Attraction to Stationary States   13
1.1   Free d'Alembert Equation   13
1.2   A String Coupled to a Nonlinear Oscillator   14
1.3   String Coupled to Several Nonlinear Oscillators   28
1.4   Space-Localized Nonlinearity   43
1.5   Wave–Particle System   56
1.6   Maxwell–Lorentz Equations: Radiation Damping   66
1.7   Wave Equations with Concentrated Nonlinearities   68
1.8   Comparison with Dissipative Systems   76

2   Global Attraction to Solitons   77
2.1   Translation-Invariant Wave–Particle System   77
2.2   The Case of Weak Coupling   89

3   Global Attraction to Stationary Orbits   91
3.1   Nonlinear Klein–Gordon Equation   91
3.2   Generalizations and Open Questions   94
3.3   Omega-Limit Trajectories   95
3.4   Limiting Absorption Principle   96
3.5   A Nonlinear Analog of Kato's Theorem   99
3.6   Splitting into Dispersive and Bound Components   102
3.7   Omega-Compactness   103
3.8   Reduction of Spectrum to Spectral Gap   104
3.9   Reduction of Spectrum to a Single Point   105
3.10 On the Nonlinear Radiative Mechanism   107
3.11 Conjecture on Attractors of G-Invariant PDEs   111

4   Asymptotic Stability of Stationary Orbits and Solitons   114
4.1   Orthogonal Projection   114
4.2   Symplectic Projection   116
4.3   Generalizations and Applications   120
4.4   Further Generalizations   122
4.5   The 1D Schrödinger Equation Coupled to an Oscillator   124

5   Adiabatic Effective Dynamics of Solitons   166
5.1   Solitons in Slowly Varying External Potentials   166
5.2   Mass–Energy Equivalence   168

6   Numerical Simulation of Solitons   170
6.1   Kinks of Relativistic Equations   170
6.2   Numerical Observation of Soliton Asymptotics   174
6.3   Adiabatic Effective Dynamics of Relativistic Solitons   174

7   Dispersive Decay   178
7.1   The Schrödinger and Klein–Gordon Equations   178
7.2   Decay $L^1 - L^\infty$ for 3D Schrödinger Equations   180

8   Attractors and Quantum Mechanics   192
8.1   Bohr's Postulates   192
8.2   On Dynamical Interpretation of Quantum Jumps   194
8.3   Bohr's Postulates via Perturbation Theory   197
8.4   Conclusion   198

Bibliography   200
Index   212

Preface

We present the theory of attractors of nonlinear Hamiltonian partial differential equations in infinite space. This is a new branch of the theory of attractors of PDEs initiated by one of the authors in 1990. This theory differs significantly from the case of dissipative systems. In particular, this theory has no analog for finite-dimensional Hamiltonian equations.

This book is the first monographic publication in this direction. Included are results on global attraction to stationary states, to solitons, and to stationary orbits; results on adiabatic effective dynamics of solitons and their asymptotic stability; and results on dispersive decay for linear Hamiltonian PDEs. The obtained results are generalized in the formulation of a new mathematical conjecture on global attractors of G-invariant nonlinear Hamiltonian partial differential equations.

We also describe the results of numerical simulations.

In conclusion, we discuss possible relations of this theory with the problem of mathematical interpretation of Bohr's transitions between quantum stationary states. The book is intended for

1. graduate and postgraduate students working with partial differential equations;
2. lecturers on PDEs;
3. mathematicians working in PDEs, mathematical physics, and mathematical problems of quantum theory.

On the Required Knowledge

All proofs are self-contained, and their overwhelming parts rely on traditional methods of analysis: ODEs, general theory of Hilbert and Banach spaces, distributions and their Fourier transform, Sobolev spaces, and definitions of Lie groups and Lie algebra and of their representations.

The key points of the proofs rely on a novel application of subtle methods of harmonic analysis: the Wiener Tauberian theorem, the Titchmarsh theorem on convolution, the theory of multipliers in the space of quasimeasures, and others. The applications are explained in detail and with exact references to the corresponding textbooks.

Acknowledgments

The authors express their deep gratitude to H. Spohn and B. Vainberg for long-time collaboration on attractors of Hamiltonian PDEs and to A. Shnirelman for useful long-term discussions. We are also grateful to A. Comech and V. Imaykin for their collaboration.

The authors are indebted to the Faculty of Mathematics of Vienna University and the Institute for the Information Transmission Problems of the Russian Academy of Sciences for providing congenial facilities for the work.

The authors are grateful to the Max Planck Institute for Mathematical Sciences (Leipzig) and to the München Technical University for their hospitality.

The work was supported in part by the Department of Mechanics and Mathematics of Moscow State University, by the Alexander von Humboldt Foundation, by the Austrian Science Fund (FWF) (projects P28152 and P34177), and by grants from the Deutsche Forschungsgemeinschaft and the Russian Fund for Basic Research.

Introduction

This monograph presents the theory of global attractors and of the long-time behavior of solutions of nonlinear Hamiltonian partial differential equations in infinite space. This theory was initiated by one of the authors in 1990, and it has been developed in collaboration with H. Spohn since 1995 and with V. S. Buslaev, A. Comech, V. Imaikin, E. Kopylova, D. Stuart, and B. Vainberg since 2005. The theory resulted, in particular, in the first rigorous solution of the problem of radiation damping in classical electrodynamics and in the first rigorous model of Bohr's transitions between quantum stationary states. This progress became possible due to novel application of subtle methods of the Wiener Tauberian theorem and the Titchmarsh convolution theorem.

The theory of attractors for nonlinear PDEs began in Landau's famous 1944 paper [22], where he proposed the first mathematical interpretation of the onset of turbulence as the growth of the dimension of attractors of the Navier–Stokes equations when the Reynolds number increases.

The foundation for the corresponding mathematical theory was laid in 1951 by Hopf, who first established the existence of global solutions of the 3D Navier–Stokes equations [5]. He introduced the *method of compactness*, which is a nonlinear version of Faedo–Galerkin approximations. This method is based on a priori estimates and Sobolev embedding theorems and has had an essential influence on the development of the theory of nonlinear PDEs (see [2, 3, 12]).

The modern development of the theory of global attractors for *dissipative PDEs*, that is, PDEs with friction, originated in 1975–1985 in publications by J. Ball, C. Foias, J. M. Ghidaglia, J. K. Hale, D. Henry, and R. Temam and was developed further by M. I. Vishik, A. V. Babin, V. V. Chepyzhov, A. Haraux, A. A. Ilyin, A. Miranville, V. Pata, E. Titi, S. Zelik, and others. An essential part of the theory up to 2000 was covered in the monographs [16]–[23].

One of the central subjects of research in this theory is the global attractor of all bounded subsets of the corresponding Banach phase space. Typically, this attractor is a submanifold connecting stationary states, which is an analog of separatrices. Each single point also attracts to this submanifold and eventually converges to one of stationary states,

$$\psi(x,t) \to S(x), \qquad t \to +\infty, \tag{1}$$

where the convergence holds in appropriate norm on the Banach phase space. In particular, the *relaxation to an equilibrium regime* in chemical reactions is due to energy dissipation.

The results obtained concern a wide class of nonlinear *dissipative* PDEs, including fundamental equations of applied and mathematical physics: the Navier–Stokes equations, nonlinear parabolic equations, reaction–diffusion equations, wave equations with friction, integro-differential equations, equations with delay, equations with memory, and so on. The techniques of functional analysis of nonlinear PDEs were developed for the study of the structure of different types of attractors; their smoothness and their fractal and Hausdorff dimensions; and their dependence on parameters, on averaging, and so on.

The development of a similar theory for *Hamiltonian PDEs* seemed at first to be unmotivated and even impossible in view of energy conservation and time reversal for these equations. However, it turned out that such a theory is possible, and its development was inspired by the problem of mathematical interpretation of basic postulates of quantum theory. These relations to quantum theory are discussed in the final chapter (Chapter 8). More details can be found in [214].

Results obtained between 1990 and 2020 suggest that long-time global attraction to a finite-dimensional submanifold in the corresponding Hilbert phase space is, in fact, a typical feature for nonlinear Hamiltonian PDEs in infinite space. These results are presented in our monograph.

For Hamiltonian PDEs in infinite space, the theory of attractors differs significantly from the case of dissipative systems, where the global attraction to stationary states is caused by an energy dissipation that is due to friction. For Hamiltonian PDEs the friction and energy dissipation are absent, and the global attraction is caused by radiation that irreversibly carries energy to infinity. This peculiarity required novel tools for analysis of nonlinear Hamiltonian PDEs, which are presented in this monograph.

Let us note, however, that this theory is only at an initial stage of its development and cannot be compared with the theory of attractors of dissipative PDEs with regard to richness and diversity of results.

The modern development of the theory of nonlinear Hamiltonian PDEs dates back to K. Jörgens [7], who first established the existence of global solutions for nonlinear wave equations of the form

$$\ddot{\psi}(x,t) = \Delta\psi(x,t) + f(\psi(x,t)), \qquad x \in \mathbb{R}^n, \qquad (2)$$

by developing the Hopf method of compactness. Subsequent studies of the well-posedness for nonlinear PDEs were presented by J.-L. Lions [12] and by T. Cazenave and A. Haraux [2, 3].

The first results on *long-time asymptotics* for *linear hyperbolic PDEs* in infinite space were established in the scattering theory by P. D. Lax, C. S. Morawetz, and R. S. Phillips for the wave equation in the exterior of a star-shaped obstacle [31]. This is the *local energy decay*: for any finite $R > 0$,

$$\int\limits_{|x|<R} [|\dot{\psi}(x,t)|^2 + |\nabla\psi(x,t)|^2 + |\psi(x,t)|^2]dx \to 0, \qquad t \to \pm\infty. \qquad (3)$$

This decay means that the energy escapes each bounded region for large times. For general linear hyperbolic PDEs and systems, similar local decay was established by B. R. Vainberg [37]. The extension of this decay to *nonlinear Hamiltonian PDEs* was established first by I. Segal, C. S. Morawetz, and W. Strauss [32]–[36]. In these papers the local energy decay (3) was proved for solutions of equations (2) with small initial data in the case of *defocusing nonlinearities* similar to

$$f(\psi) = -m^2\psi - \varkappa|\psi|^{p-1}\psi, \qquad (4)$$

where $m^2 \geq 0, \varkappa > 0$, and $p > 1$. Moreover, in these articles the corresponding nonlinear wave operators and scattering operators are constructed. In [77, 78], W. Strauss established the completeness of the scattering for small solutions of more general equations.

For convenience, characteristic properties of all finite-energy solutions of an equation will be referred to as *global* to distinguish them from the corresponding *local* properties of the solutions with initial data sufficiently close to an attractor. Note that global attraction to a (proper) attractor is impossible for finite-dimensional Hamiltonian systems because of energy conservation. All the aforementioned results [32]–[36] on local energy decay (3) for nonlinear Hamiltonian PDEs mean that the corresponding *local attractor* of solutions with small initial states consists of only the zero point.

Theory of global attractors The first results on *global attractors* for nonlinear Hamiltonian PDEs were obtained by one of the present authors in 1991–1995 for 1D equations [40, 41, 42] and were extended to multidimensional equations in 1995–2020 in collaboration with A. Comech, V. S. Buslaev, E. Kopylova, H. Spohn, D. Stuart, B. R. Vainberg, and others. These results were obtained from an analysis of the irreversible energy radiation to infinity, which plays the role of dissipation. This progress was achieved by a novel application of subtle methods of harmonic analysis: the Wiener Tauberian theorem, the Titchmarsh convolution theorem, the new theory of multipliers in the space of quasimeasures, and other methods.

The questions of asymptotic stability required the use of the stationary scattering theory of Agmon, Jensen, and Kato [171, 183] and of the eigenfunction expansion for non-selfadjoint Hamiltonian operators [137, 138] based on M. G. Krein's theory of J-selfadjoint operators.

One of the key observations is that the results obtained so far indicate a certain dependence of long-time asymptotics of solutions on the symmetry group of the equation. For example, it may be the trivial group $G = \{e\}$, or the group of translations $G = \mathbb{R}^n$, or the unitary group $G = U(1)$, or the orthogonal group $SO(3)$. This observation suggests general conjecture for nonlinear Hamiltonian *autonomous* PDEs of type

$$\dot{\Psi}(t) = F(\Psi(t)), \qquad t \in \mathbb{R}, \tag{5}$$

with a Lie symmetry group G, which acts on the Hilbert or Banach phase space $\mathcal{E}$ of the equation via a representation T.

Conjecture A (On attractors) *For* **generic** *nonlinear Hamiltonian PDEs (5) with a Lie symmetry group G, any finite-energy solution admits the asymptotics*

$$\Psi(t) \sim e^{\hat{\lambda}_{\pm} t} \Psi_{\pm}, \qquad t \to \pm\infty \tag{6}$$

in the appropriate topology of the phase space $\mathcal{E}$.

Here $\hat{\lambda}_{\pm} = T'(e)\lambda_{\pm}$, where $\lambda_{\pm}$ belong to the corresponding Lie algebra $\mathfrak{g}$, while the $\Psi_{\pm}(x)$ are some *limiting amplitudes* depending on the trajectory $\Psi(x,t)$ considered. Both pairs $(\Psi_+, \hat{\lambda}_+)$ and $(\Psi_-, \hat{\lambda}_-)$ are solutions of the corresponding *nonlinear eigenvalue problem* (3.11.5); see more details in Section 3.11.

Let us specify the asymptotics (6) for the four symmetry groups mentioned above.

1. Equations with trivial symmetry group $G = \{e\}$ For such *generic* equations, the conjecture (6) means *global attraction to stationary states*

$$\psi(x,t) \to S_{\pm}(x), \qquad t \to \pm\infty, \tag{7}$$

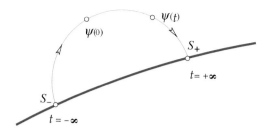

Figure 1 Convergence to stationary states.

as is illustrated in Figure 1. Here the states $S_\pm(x)$ depend on the trajectory $\psi(x,t)$ under consideration, and the convergence holds in local seminorms of type $L^2(|x| < R)$ with any $R > 0$. This convergence cannot hold in global norms (i.e., in norms corresponding to $R = \infty$) due to energy conservation. The asymptotics (7) can be symbolically written as the transitions

$$S_- \mapsto S_+, \tag{8}$$

which can be considered as the mathematical model of Bohr's quantum jumps (8.1.1).

Such an attraction was established in [40]–[52] for a variety of model equations: (1) for a string coupled to nonlinear oscillators, (2) for a 3D wave equation coupled to a charged particle and for the Maxwell–Lorentz equations, and also (3) for wave equations and Dirac and Klein–Gordon equations with concentrated nonlinearities.

All proofs rely on the bounds for radiation which irreversibly carries energy to infinity. The proofs of global attraction in [44, 45] rely on a novel application of the Wiener Tauberian theorem [15], which provides the relaxation of the acceleration of the particle

$$\ddot{q}(t) \to 0, \qquad t \to \pm\infty \tag{9}$$

under the *Wiener condition* (1.5.13) on the particle charge density. These results gave the first rigorous proof of *radiation damping* (9) in classical electrodynamics, which has been an open problem for about 100 years.

The results of [40]–[44] and [50] are presented with detail in Chapter 1.

In all problems considered here, the convergence (7) implies by the Fatou theorem the inequality

$$\mathcal{H}(S_\pm) \leq \mathcal{H}(Y(t)) \equiv \text{const}, \quad t \in \mathbb{R}, \tag{10}$$

where $\mathcal{H}$ is the corresponding Hamiltonian (energy) functional. This inequality is an analog of the well-known property of the weak convergence in the Hilbert and Banach spaces. Simple examples show that strong inequality in (10) is possible, which means the irreversible scattering of energy to infinity.

Example 1 The d'Alembert waves In particular, the asymptotics (7) and the strong inequality (10) can easily be demonstrated for the d'Alembert equation with general solution $\psi(x,t) = f(x-t) + g(x+t)$. Namely, the local convergence $\psi(\cdot,t) \to 0$ in $L^2_{\mathrm{loc}}(\mathbb{R})$ obviously holds for all $f, g \in L^2(\mathbb{R})$. On the other hand, the convergence to zero in the global norm of $L^2(\mathbb{R})$ obviously fails if $f(x) \not\equiv 0$ or $g(x) \not\equiv 0$.

Example 2 Nonlinear strong Huygens principle Similarly, a solution of the 3D wave equation with unit speed of propagation is concentrated in spherical layers $|t| - R < |x| < |t| + R$ if the initial data have support in the ball $|x| \leq R$. Therefore, the solution converges to zero in $L^2_{\mathrm{loc}}(\mathbb{R}^3)$ as $t \to \pm\infty$, although its energy remains constant. This also illustrates the strong inequality in (10). This convergence corresponds to the well-known *strong Huygens principle* in optics and acoustics (see [1]). Thus, global attraction to stationary states (7) is a generalization of the strong Huygens principle to nonlinear equations. The difference is that for the linear wave equation the limit is always zero, while for nonlinear equations the limit can be any stationary solution.

2. Equations with the symmetry group of translations $G = \mathbb{R}^n$ Let us consider, as an example, the case of the simplest representation

$$[T(a)\psi](x) := \psi(x - a), \qquad x \in \mathbb{R}^n \tag{11}$$

for $a \in \mathbb{R}^n$. Then the asymptotics (6) means *global attraction to solitons* (traveling waves)

$$\psi(x,t) \sim \psi_\pm(x - v_\pm t), \qquad t \to \pm\infty, \tag{12}$$

where the asymptotics holds in local seminorms of type $L^2(|x - v_\pm t| < R)$ with any $R > 0$, that is, *in the comoving frame of reference*.

Such soliton asymptotics was proved first for *integrable equations* (Korteweg–de Vries equation (KdV), etc.); see [53, 59]. Moreover, for the Korteweg–de Vries equation, more accurate soliton asymptotics in *global norms* with several solitons were first discovered by M. Kruskal and N. J. Zabuzhsky in 1965 by numerical simulation: it is the decay to solitons

$$\psi(x,t) \sim \sum_k \psi_\pm(x - v_\pm^k t) + w_\pm(x,t), \qquad t \to \pm\infty, \tag{13}$$

where $w_\pm$ are some dispersive waves. A trivial example is provided by the d'Alembert equation $\ddot{\psi}(x,t) = \psi''(x,t)$, for which any solution reads $\psi(x,t) = f(x-t) + g(x+t)$.

Later on, such asymptotics were proved by the method of the *inverse scattering problem* for nonlinear *integrable* Hamiltonian translation-invariant equations (KdV, etc.) in the works of M. J. Ablowitz, H. Segur, W. Eckhaus, A. van Harten, and others [53, 59].

For *nonintegrable* equations the global attraction to solitons (12) was established for the first time in [54]–[57] for translation-invariant systems of the wave and Maxwell equations coupled to a charged relativistic particle. The result of [55] gives the first rigorous proof of the *radiation damping* for the translation-invariant system of classical electrodynamics.

The proofs in [54] and [55] rely on a canonical transformation to the comoving frame and variational properties of solitons, as well as on the relaxation of the acceleration (9) under the Wiener condition for the particle charge density.

The multisoliton asymptotics (13) for *nonintegrable equations* were observed numerically in [58] in the case of 1D *relativistic* nonlinear wave equations.

The results of [54] and [58] are presented with details in Chapters 2 and 6, respectively.

3. Equations with the unitary symmetry group $G = U(1)$ Let us consider for example the case of the simplest representation

$$[T(e^{i\theta})\psi](x) := e^{i\theta}\psi(x), \qquad x \in \mathbb{R}^n \tag{14}$$

for $\theta \in R$. Then the asymptotics (6) means the *single-frequency asymptotics*

$$\psi(x,t) \sim \psi_\pm(x)e^{-i\omega_\pm t}, \qquad t \to \pm\infty, \tag{15}$$

where $\omega_\pm \in \mathbb{R}$.

Example 3 For the case of the coupled Maxwell–Schrödinger equations (8.2.1) with the symmetry group $U(1)$, the conjecture (6) reduces to the asymptotics (8.2.8) similar to (15).

The asymptotics (15) also means the global attraction to the solitary manifold formed by all *stationary orbits* which are solutions of type $\psi_\omega(x)e^{-i\omega t}$. The asymptotics are expected in the local seminorms $L^2(|x| < R)$ with any $R > 0$. The global attractor is a smooth manifold formed by the circles which are the orbits of the action of the symmetry group $U(1)$ (see Figure 2).

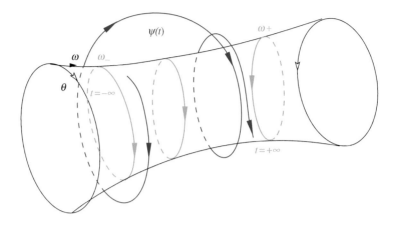

Figure 2 Convergence to stationary orbits.

Such an attraction *in local seminorms* $L^2(|x| < R)$ was proved (1) in [61]–[67] for the Klein–Gordon and Dirac equations coupled to a $U(1)$-invariant nonlinear oscillator; (2) in [60], for discrete approximations of such coupled systems, i.e., for the corresponding difference schemes; and (3) in [69]–[71] for the wave, Klein–Gordon, and Dirac equations with concentrated nonlinearities. More precisely, we have proved global attraction to the *solitary manifold* of all stationary orbits, though global attraction to particular stationary orbits, with fixed $\omega_\pm$, is still an open problem.

All these results were proved under the assumption that the equations are "strictly nonlinear." For linear equations, the global attraction obviously fails if the discrete spectrum consists of at least two different eigenvalues.

The proofs of these results rely on (1) the concept of omega-limit trajectory, (2) a nonlinear analog of the Kato theorem on the absence of embedded eigenvalues, (3) new theory of multipliers in the space of quasimeasures, and (4) novel application of the Titchmarsh convolution theorem. The results of [62]–[64] are presented with details in Chapter 3.

Existence and orbital stability of stationary orbits The existence of solutions $e^{\hat{\lambda}t}\Psi$ (*stationary G-orbits*) for G-invariant nonlinear wave equations (2) in the cases $G = U(1)$ and $G = \mathbb{R}^n$ was extensively studied in the 1960s–1980s. The most general results were obtained by W. Strauss, H. Berestycki, and P.-L. Lions [24, 25, 30]. M. Esteban, V. Georgiev, and E. Séré constructed in [27] stationary orbits for the relativistic nonlinear Maxwell–Dirac system (8.2.7) and for the Klein–Gordon–Dirac system. The key role in these papers was played by the Lusternik–Schnirelmann theory of critical points [28, 29].

In [26] G. M. Coclite and V. Georgiev constructed stationary orbits for the nonlinear Maxwell–Schrödinger system with the external Coulomb potential.

General theory of *orbital stability* of stationary G-orbits was developed by M. Grillakis, J. Shatah, and W. Strauss in [100, 101].

4. Equations with the orthogonal symmetry group $G = SO(3)$ For such generic equations, the asymptotics (6) means that

$$\psi(x,t) \sim e^{-i\hat{\Omega}_\pm t}\psi_\pm(x), \qquad t \to \pm\infty, \tag{16}$$

where $\hat{\Omega}_\pm$ are suitable representations of real skew-symmetric 3×3 matrices $\Omega_\pm \in \mathfrak{so}(3)$. This means that global attraction to "stationary $SO(3)$-orbits" occurs. Such asymptotics are proved in [88] for the Maxwell–Lorentz equations with rotating particle.

Generic equations Let us emphasize that, for example, we are conjecturing asymptotics (15) for *generic $U(1)$-invariant equations*. This means that the long-time behavior of solutions may be quite different for $U(1)$-invariant equations of "positive codimension." In particular, for solutions of the linear Schrödinger equation

$$i\dot{\psi}(x,t) = -\Delta\psi(x,t) + V(x)\psi(x,t), \qquad x \in \mathbb{R}^n,$$

the asymptotics (15) generally fail. Namely, any finite-energy solution admits the spectral representation

$$\psi(x,t) = \sum C_k \psi_k(x)e^{-i\omega_k t} + \int\limits_0^\infty C(\omega)\psi(\omega,x)e^{-i\omega t}d\omega,$$

where ψ_k and $\psi(\omega,\cdot)$ are the corresponding eigenfunctions of the discrete and continuous spectrum, respectively. The last integral is a dispersive wave, which decays to zero in the norms $L^2(|x| < R)$ with any $R > 0$ (under appropriate conditions on the potential $V(x)$). Correspondingly, the attractor is the linear span of the eigenfunctions ψ_k. Thus, the long-time asymptotics does not reduce to a single term like (15), so the linear case is degenerate in this sense. Note that all our results [61]–[67] are established for a *strictly nonlinear case* (see the condition (3.1.16)), which eliminates linear equations.

Higher symmetry groups For more sophisticated symmetry groups $G = U(N)$, the asymptotics (6) mean the global attraction to N-frequency trajectories, which can be quasi-periodic. In particular, the symmetry groups $SU(2)$, $SU(3)$, and others were suggested in 1961 by M. Gell-Mann and Y. Ne'eman for strong interaction of baryons [222, 224]. This theory provides empirical evidence for the asymptotics (6), see Section 3.11.

On relations with Soffer's conjectures Note that our conjecture (6) specifies the concept of *localized solution/coherent structures* from the "Grande Conjecture" and the "Petite Conjecture" of A. Soffer (see [161], p. 460) in the context of the Banach spaces. The Grande Conjecture was proved in [47] for the case of a 1D wave equation coupled to a nonlinear oscillator (1.2.1). Moreover, suitable versions of the Grande Conjecture were also proved in [57, 88] for the 3D wave and Klein–Gordon and Maxwell equations coupled to a relativistic particle with sufficiently small charge (2.2.1) (see Remark 2.2.1). Finally, for any matrix symmetry group G, the asymptotics (6) corresponds to the Petite Conjecture since then the localized solutions $e^{g \pm t} \psi_{\pm}(x)$ are quasi-periodic.

In this book we present available results on the global attraction (7)–(16) and related numerical experiments. Moreover, we survey the results on asymptotic stability of solitons and their adiabatic effective dynamics, on the dispersive decay and relations to quantum mechanics.

Asymptotic stability of solitons More precisely, we should phrase "asymptotic stability of solitary manifolds," which means a local attraction, i.e., for states sufficiently close to the manifold. There is a huge body of literature on this subject. In Chapter 4 we review the results on such local attraction that were developed in a series of articles [162]–[170] by V.S. Buslaev, G. Perelman, A. Soffer, D. Stuart, C. Sulem, T. P. Tsai, M. Weinstein, H. T. Yau, and others.

The crucial peculiarity of this attraction is the instability of the dynamics *along the solitary manifold*. This follows directly from the fact that solitons move with different speeds and therefore run away for large times. Analytically, this instability is caused by the presence of the eigenvalue $\lambda = 0$ in the spectrum of the generator of linearized dynamics. Namely, the tangent vectors to the solitary manifold are eigenvectors and associated vectors of the generator. They correspond to zero eigenvalue. Respectively, the Lyapunov theory is not applicable to this case.

This is why in the articles [162]–[169] an original strategy was developed for proving asymptotic stability of solitary manifolds. This strategy allows one to separate the unstable motion along the solitary manifold and the attraction in transversal directions to this manifold.

This approach relies on (1) a special projection of a trajectory onto the solitary manifold, (2) modulation equations for parameters of the projection, and (3) time decay of the transversal component. It is a far-reaching development of the Lyapunov stability theory.

Adiabatic effective dynamics of solitons In Chapter 5 we describe adiabatic effective dynamics for solitons in slowly varying external potentials, when the corresponding external force is small. The existence of solitons and the global attraction to solitons (12) are typical features of translation-invariant systems. However, if the deviation of a system from translational invariance is in some sense small, the system can admit solutions that are close forever to solitons with time-dependent parameters (e.g., velocity). Moreover, in some cases it is possible to identify an "effective dynamics" that describes the evolution of these parameters.

We present without proofs the results of [84] and [85] on *adiabatic effective dynamics* (5.1.5), (5.1.6) for the wave–particle system (1.5.1)–(1.5.2) and the Maxwell–Lorentz system (1.6.1), respectively, in the case of slowly varying external potentials. We also discuss the related *mass–energy equivalence*.

In Chapter 6 we present results of numerical simulation of soliton asymptotics and on the corresponding effective dynamics for relativistic equations.

Dispersive decay In Chapter 7 we give (1) a brief survey of basic results on the dispersive decay and (2) a new short and simplified proof of the fundamental results on the $L^1 \to L^\infty$ dispersive decay established by J.-L. Journé, A. Soffer, and C. D. Sogge in [184] for the Schrödinger equation (7.1.2) with $n \geq 3$.

The dispersive decay of the corresponding linearized equations plays the key role in all results on long-time asymptotics for nonlinear Hamiltonian PDEs. One of the first fundamental results on the dispersive decay is the local energy decay (3) established in [31].

Relations to quantum mechanics In the final chapter, Chapter 8, we discuss possible relationships between the theory of attractors of Hamiltonian nonlinear equations and quantum mechanics. The theory of global attraction (15) was inspired by postulates of N. Bohr on transitions to *quantum stationary states* and by Schrödinger's definition of these quantum stationary states as solutions of type $\psi(x,t) = \psi(x)e^{-i\omega t}$ (see Chapter 8 for details). Our results confirm such attraction for *generic* $U(1)$-invariant nonlinear equations of type (3.1.1) and (3.2.1)–(3.2.3). However, for the semiclassical self-consistent Maxwell–

Schrödinger system of quantum mechanics (8.2.1), this attraction is still an open, challenging problem.

On related surveys In conclusion, let us mention the related surveys in this area [8, 11, 46]. The results on asymptotic stability of solitary manifolds were described in detail in [124] for linear equations coupled to a particle and in [144] for the relativistic Ginzburg–Landau equations.

1

Global Attraction to Stationary States

In this chapter we present with details the results on global attraction to stationary states (7) for nonlinear Hamiltonian PDEs in infinite space.

In Section 1.2 we present the first result of this type obtained in [40, 41] for a 1D wave equation coupled to one nonlinear oscillator ("the Lamb system"). The second result [42] for a 1D wave equation coupled to several nonlinear oscillators is presented in Section 1.3, and the third result – for a 1D wave equation coupled to a "continuum of nonlinear oscillators" – is presented in Section 1.4. The proof of the last result relies on calculation of energy radiation to infinity and uses the concept of omega-limit trajectories.

In Sections 1.5 and 1.6 are presented the results [44] and [45], which concern, respectively, 3D wave equations and Maxwell's equations coupled to a charged relativistic particle with density of charge satisfying the Wiener condition. In particular, the radiation damping in classical electrodynamics is rigorously proved for the first time. The proofs rely on calculation of energy radiation to infinity and use the Wiener Tauberian theorem.

Section 1.7 concerns the result [50] on a 3D wave equation with concentrated nonlinearities. The key step in the proof is an investigation of a nonlinear integro-differential equation.

1.1 Free d'Alembert Equation

The global attraction (7) can easily be demonstrated using the trivial (but instructive) example of the d'Alembert equation

$$\ddot{\psi}(x,t) = \psi''(x,t), \qquad x \in \mathbb{R}, \tag{1.1.1}$$

where $\dot{\psi} := \frac{\partial \psi}{\partial t}$, $\psi' := \frac{\partial \psi}{\partial x}$. All derivatives here and below are understood in the sense of distributions. This equation is formally equivalent to the Hamiltonian system

$$\dot{\psi}(t) = D_\pi \mathcal{H}, \quad \dot{\pi}(t) = -D_\psi \mathcal{H} \qquad (1.1.2)$$

with Hamiltonian

$$\mathcal{H}(\psi, \pi) = \frac{1}{2} \int [|\pi(x)|^2 + |\psi'(x)|^2] \, dx,$$

$$(\psi, \pi) \in \mathcal{E} := H_c^1(\mathbb{R}) \oplus [L^2(\mathbb{R}) \cap L^1(\mathbb{R})], \qquad (1.1.3)$$

where $H_c^1(\mathbb{R})$ is the Hilbert space of continuous functions $\psi(x)$ with finite norm

$$\|\psi\|_{H_c^1(\mathbb{R})} := \|\psi'\|_{L^2(\mathbb{R})} + |\psi(0)|. \qquad (1.1.4)$$

Let us consider solutions $(\psi(x,t), \pi(x,t))$ of (1.1.2) with initial states $(\psi(x,0), \pi(x,0)) = (\psi_0(x), \pi_0(x)) \in \mathcal{E}$. Let us assume, moreover, that

$$\psi_0(x) \to C_\pm, \quad x \to \pm\infty. \qquad (1.1.5)$$

For such initial data the d'Alembert formula gives

$$\psi(x,t) = \frac{\psi_0(x+t) + \psi_0(x-t)}{2} + \frac{1}{2} \int_{x-t}^{x+t} \pi_0(y) dy$$

$$\xrightarrow[t \to \pm\infty]{} S_\pm(x) = \frac{C_+ + C}{2} \pm \frac{1}{2} \int_{-\infty}^{\infty} \pi_0(y) dy, \qquad (1.1.6)$$

where the convergence is uniform on every finite interval $|x| < R$. Moreover,

$$\dot{\psi}(x,t) = \frac{\psi_0'(x+t) - \psi_0'(x-t)}{2} + \frac{\pi_0(x+t) + \pi_0(x-t)}{2} \to 0, \quad t \to \pm\infty, \tag{1.1.7}$$

where the convergence holds in $L^2(-R, R)$ for each $R > 0$. Thus, the set of stationary states $(\psi(x), \pi(x)) = (C, 0)$, where $C \in \mathbb{R}$ is any constant, is an attractor. Note that for positive and negative times the limits (1.1.6) may be different.

1.2 A String Coupled to a Nonlinear Oscillator

In this section we present the first results on global attraction to stationary states (7) for nonlinear Hamiltonian PDEs obtained in [40, 41] (and developed in [47]) for the *nonlinear Lamb system* with a point nonlinearity:

$$\begin{cases} \ddot{\psi}(x,t) = \psi''(x,t), & x \in \mathbb{R} \setminus \{0\}, \\ m\ddot{y}(t) = F(y(t)) + \psi'(+0,t) - \psi'(-0,t); \ y(t) \equiv \psi(0,t), \end{cases} \quad (1.2.1)$$

where $m > 0$. Solutions $\psi(x,t)$ take the values in $\mathbb{R}^d$ with $d \geq 1$. This system can formally be written as the nonlinear wave equation

$$(1 + m\delta(x))\ddot{\psi}(x,t) = \psi''(x,t) + \delta(x)F(\psi(0,t)), \qquad x \in \mathbb{R}. \quad (1.2.2)$$

The problem (1.2.1) describes small crosswise oscillations of an infinite string stretched parallel to the x-axis; a particle of mass $m > 0$ is attached to the string at the point $x = 0$; $F(y)$ is an external (nonlinear) force perpendicular to the string; the force subjects the particle (see Figure 1.1).

The system (1.2.1) has been introduced originally by H. Lamb [51] in the linear case when $F(y) = -\omega^2 y$. The Lamb system with nonlinear force $F(y)$ has been considered in [39], where the questions of irreversibility and nonrecurrence were discussed. The system was studied further in [40, 41, 47], where the global attraction to stationary states has been established for the first time, and in [38], where metastable regimes were studied for the stochastic Lamb system with white noise.

The Lamb system (1.2.1) is a simplest nontrivial nonlinear time-reversible infinite-dimensional Hamiltonian system allowing an effective analysis of various questions.

Our main results for this system are as follows. Here we establish the existence of a finite-dimensional global attractor and establish the nonlinear scattering:

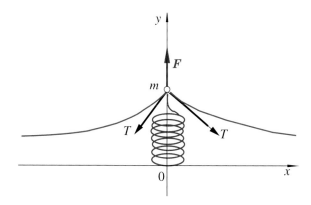

Figure 1.1 String coupled to an oscillator.

> *Each finite-energy solution decays in long time limits to a sum*
> *of a stationary state and a dispersive wave.*

The asymptotics holds in global energy norm. Moreover, in [48, 49], we have established the asymptotic completeness of the corresponding nonlinear scattering operators.

We consider the Cauchy problem for the system (1.2.1) with the initial conditions

$$\psi|_{t=0} = \psi_0(x), \qquad \dot{\psi}|_{t=0} = \pi_0(x), \qquad \dot{y}|_{t=0} = p_0. \tag{1.2.3}$$

Denote $Y(t) = (\psi(x,t), \dot{\psi}(x,t), \dot{y}(t))$. Then the Cauchy problem (1.2.1), (1.2.3) can be written as

$$\dot{Y}(t) = \mathbf{F}(Y(t)) \text{ for } t \in \mathbb{R}, \quad Y(0) = Y_0, \tag{1.2.4}$$

where $Y_0 = (\psi_0, \pi_0, p_0)$ and

$$\mathbf{F}(Y(t)) = (\dot{\psi}(\cdot,t), \psi''(x,t)|_{x \neq 0}, F(\psi(0,t)) + \psi'(+0,t) - \psi'(-0,t)).$$

An exact statement of the Cauchy problem will be formulated in the next section.

We will establish the scattering asymptotics

$$Y(t) \sim S_{\pm} + \tilde{W}(t)\Psi_{\pm}, \qquad t \to \pm\infty, \tag{1.2.5}$$

where $S_{\pm}$ are some stationary states of the system (1.2.1), $\tilde{W}(t)$ is the dynamical group of the free wave equation, and $\Psi_{\pm} \in \mathcal{E}$ are the corresponding scattering states. The asymptotics (1.2.5) holds if the following limits exist:

$$\psi_0^+ := \lim_{x \to +\infty} \psi_0(x), \quad \psi_0^- := \lim_{x \to -\infty} \psi_0(x), \quad I_0 := \int_{-\infty}^{\infty} \pi_0(y) dy. \tag{1.2.6}$$

1.2.1 Hilbert Phase Space and Dynamics

Let us introduce a Hilbert phase space $\mathcal{E}$ of finite-energy states for the system (1.2.1). Denote by $\| \cdot \|$ resp. $\| \cdot \|_R$ the norm in the Hilbert space $L^2 := L^2(\mathbb{R}, \mathbb{R}^d)$ resp. $L^2([-R, R], \mathbb{R}^d)$ and by $E_c := H_c^1(\mathbb{R}) \otimes \mathbb{R}^d$, where $H_c^1(\mathbb{R})$ is the Hilbert space with the norm (1.1.4).

Definition 1.2.1 (i) $\mathcal{E}$ is the Hilbert space of triples $(\psi(x), \pi(x), p) \in E_c \oplus L^2 \oplus \mathbb{R}^d$ with finite energy norm

$$\|(\psi, \pi, p)\|_{\mathcal{E}} = \|\psi\|_{E_c} + \|\pi\| + |p| = \|\psi'\| + |\psi(0)| + \|\pi\| + |p|. \tag{1.2.7}$$

(ii) $\mathcal{E}_F$ is the space $\mathcal{E}$ endowed with the topology defined by the *local energy seminorms*

$$\|(\psi,\pi,p)\|_{\mathcal{E},R} \equiv \|\psi'\|_R + |\psi(0)| + \|\pi\|_R + |p|, \quad R > 0. \qquad (1.2.8)$$

The space $\mathcal{E}_F$ is not complete, and convergence in $\mathcal{E}_F$ is equivalent to convergence in the metric

$$\text{dist}[Y_1,Y_2] = \sum_1^\infty 2^{-R} \frac{\|Y_1 - Y_2\|_{\mathcal{E},R}}{1 + \|Y_1 - Y_2\|_{\mathcal{E},R}}, \quad Y_1, Y_2 \in \mathcal{E}. \qquad (1.2.9)$$

We assume that

$$F(y) \in C^1(\mathbb{R}^d, \mathbb{R}^d), \quad F(y) = -\nabla V(y), \qquad (1.2.10)$$

$$V(y) \to +\infty, \quad y \to \infty. \qquad (1.2.11)$$

In this case the system (1.2.1) is formally Hamiltonian with the Hilbert phase space $\mathcal{E}$ and the Hamiltonian functional

$$\mathcal{H}(\psi,\pi,p) = \frac{1}{2} \int [|\pi(x)|^2 + |\psi'(x)|^2]dx + m\frac{|p|^2}{2} + V(\psi(0)) \qquad (1.2.12)$$

for $(\psi,\pi,p) \in \mathcal{E}$. We consider solutions $\psi(x,t)$ such that $Y(t) = (\psi(\cdot,t), \dot\psi(\cdot,t), \dot y(t)) \in C(\mathbb{R}, \mathcal{E})$, where $y(t) := \psi(0,t)$.

Let us discuss the definition of the Cauchy problem (1.2.1), (1.2.3) for the trajectories $Y(t) \in C(\mathbb{R}, \mathcal{E})$. At first, $\psi \in C(\mathbb{R}^2, \mathbb{R}^d)$ for $Y(t) \in C(\mathbb{R}, \mathcal{E})$. Hence, the first equation in (1.2.1) is equivalent to the d'Alembert decomposition

$$\psi(x,t) = f_\pm(x - t) + g_\pm(x + t), \quad \pm x > 0, \qquad (1.2.13)$$

where

$$f_\pm, g_\pm \in C(\mathbb{R}, \mathbb{R}^d).$$

Therefore,

$$\dot\psi(x,t) = -f'_\pm(x - t) + g'_\pm(x + t),$$

$$\psi'(x,t) = f'_\pm(x - t) + g'_\pm(x + t) \text{ for } \pm x > 0,$$

where all the derivatives are understood in the sense of distributions. The assumption $Y(t) \in C(\mathbb{R}, \mathcal{E})$ implies

$$f'_\pm, g'_\pm \in L^2_{\text{loc}}(\mathbb{R}, \mathbb{R}^d).$$

Now we explain the second equation of (1.2.1).

Definition 1.2.2 In the second equation of (1.2.1) we set

$$\psi'(0\pm,t) := f'_\pm(-t) + g'_\pm(t) \in L^2_{loc}(\mathbb{R},\mathbb{R}^d), \qquad (1.2.14)$$

while the derivative $\ddot{y}(t)$ of $y(t) \equiv \psi(0,t) \in C(\mathbb{R},\mathbb{R}^d)$ is understood in the sense of distributions.

Note that the functions $f_\pm$ and $g_\pm$ in (1.2.13) are unique up to an additive constant. Hence definition (1.2.14) is unambiguous.

Proposition 1.2.3 (cf. [41]) *Let the conditions (1.2.10), (1.2.11) hold, $m > 0$, and $Y_0 \in \mathcal{E}$. Then*
(i) The Cauchy problem (1.2.4) admits a unique solution $Y(t) \in C(\mathbb{R},\mathcal{E})$.
(ii) The map $U(t) : Y_0 \mapsto Y(t)$ is continuous in $\mathcal{E}$ and in $\mathcal{E}_F$.
(iii) The energy is conserved,

$$\mathcal{H}(Y(t)) = \text{const}, \quad t \in \mathbb{R}. \qquad (1.2.15)$$

(iv) The a priori bounds hold:

$$\sup_{t\in\mathbb{R}} \|Y(t)\|_\mathcal{E} < \infty.$$

1.2.2 Main Results

The stationary states $S = (s(x),0,0) \in \mathcal{E}$ for the system (1.2.1) are evidently determined: the set $\mathcal{S}$ of all stationary states $S \in \mathcal{E}$ is given by

$$\mathcal{S} = \{S_z = (z,0,0) : z \in Z\}, \qquad \text{where} \quad Z := \{z \in \mathbb{R}^d : F(z) = 0\}. \qquad (1.2.16)$$

The next theorem means that the set $\mathcal{S}$ is the global point attractor of the system (1.2.1) in the space $\mathcal{E}_F$.

Theorem 1.2.4 (cf. [40, 41]) *Let all assumptions of Proposition 1.2.3 hold and an initial state $Y_0 \in \mathcal{E}$. Then*
(i) The corresponding solution $Y(t) \in C(\mathbb{R},\mathcal{E})$ to the Cauchy problem (1.2.4) attracts to the set $\mathcal{S}$,

$$Y(t) \xrightarrow{\mathcal{E}_F} \mathcal{S}, \quad t \to \pm\infty \qquad (1.2.17)$$

in the metric (1.2.9). This means that

$$\text{dist}[Y(t),\mathcal{S}] := \inf_{S\in\mathcal{S}} \text{dist}[Y(t), S] \to 0, \quad t \to \pm\infty. \qquad (1.2.18)$$

(ii) Suppose additionally that the set Z is a discrete subset in $\mathbb{R}^d$. Then any solution $Y(t) \in C(\mathbb{R}, \mathcal{E})$ attracts to some stationary states $S_\pm \in \mathcal{S}$ depending on the solution,

$$Y(t) \xrightarrow{\mathcal{E}_F} S_\pm, \quad t \to \pm\infty, \tag{1.2.19}$$

as is illustrated in Figure 1.

Remarks 1.2.5 (i) The discreteness of the set Z is essential for the global attraction to stationary states (1.2.19). For example, let us consider the nonlinearity which vanishes on a C^1-submanifold of $\mathbb{R}^d$,

$$F(\psi) \equiv 0, \qquad \psi \in I. \tag{1.2.20}$$

Then, in the case $m = 0$, any smooth function $f(x - t)$ with values in I is the solution to the system (1.2.1). In particular, for $d = 1$ and $I = [-1, 1]$, we can take the function

$$\psi(x,t) = \sin \log(|x - t| + 2), \qquad (x,t) \in \mathbb{R}^2. \tag{1.2.21}$$

In this case the function $(\psi(x,t), \dot\psi(x,t), \psi(0,t)) \in C(\mathbb{R}, \mathcal{E})$ is the solution to equation (1.2.4) with $m = 0$, and for this solution the attraction to stationary states (1.2.19) obviously breaks down. On the other hand, (1.2.17) for this solution holds. For $m > 0$, similar examples can be easily constructed; see [41].

(ii) The "weak convergence" (1.2.19) and (1.2.11), (1.2.12) imply (10) by the Fatou lemma.

Furthermore, let us denote $\mathcal{E}_0 = \{(\psi, v, 0) \in \mathcal{E}\}$ and $\tilde{W}(t)(\psi, v, 0) := (W(t)(\psi, v), 0)$, where $W(t)$ is the dynamical group of free wave equation (1.1.1).

Theorem 1.2.6 ([47]) *Let all assumptions of Proposition 1.2.3 hold, and additionally, the finite limits (1.2.6) exist. Then the* **scattering asymptotics** *hold:*

$$Y(t) = S_\pm + \tilde{W}(t)\Psi_\pm + r_\pm(t), \tag{1.2.22}$$

with $S_\pm \in \mathcal{S}$, and some asymptotic states $\Psi_\pm \in \mathcal{E}_0$; the remainder is small in the **global energy norm** *(1.2.7):*

$$\|r_\pm(t)\|_{\mathcal{E}} \to 0, \quad t \to \pm\infty. \tag{1.2.23}$$

In [48, 49] the asymptotic completeness of the corresponding nonlinear scattering operator $S : \Psi_- \mapsto \Psi_+$ has been proved for equation (1.2.1) in the case $m = 0$.

1.2.3 Well-Posedness

Recall, since we need some of its constructions later, the proof of Proposition 1.2.3 from [41] in the proofs of Theorems 1.2.4 and 1.2.6.

The construction of solutions relies on the d'Alembert representation (1.2.13). For $\pm z > 0$ the functions $f_\pm(z)$ and $g_\pm(z)$ are defined by the d'Alembert formulas

$$
\begin{aligned}
f_\pm(z) &:= \frac{\psi_0(z)}{2} - \frac{1}{2} \int_0^z \pi_0(y)\, dy, \\
&\qquad\qquad\qquad\qquad\qquad \pm z > 0. \qquad (1.2.24) \\
g_\pm(z) &:= \frac{\psi_0(z)}{2} + \frac{1}{2} \int_0^z \pi_0(y)\, dy,
\end{aligned}
$$

These formulas imply that

$$
f'_\pm(z), g'_\pm(z) \in L^2(\mathbb{R}^\pm, \mathbb{R}^d), \qquad (1.2.25)
$$

since $(\psi_0, \pi_0) \in \mathcal{E}$. The *reflected outgoing waves* $f_+(z)$ for $z < 0$ and $g_-(z)$ for $z > 0$ are given by

$$
f_+(-t) := y(t) - g_+(t), \qquad g_-(t) := y(t) - f_-(-t), \qquad t > 0 \qquad (1.2.26)
$$

due to the gluing conditions $y(t) := \psi(0,t) = f_+(-t) + g_+(t) = f_-(-t) + g_-(t)$. Hence,

$$
\psi(x,t) = \begin{cases}
y(t-x) + g_+(x+t) - g_+(t-x), & 0 < x < t \\
y(t+x) + f_-(x-t) - f_-(-x-t), & -t < x < 0
\end{cases} \quad t > 0. \qquad (1.2.27)
$$

Substituting these representations into the second equation of (1.2.1), we immediately get the *reduced equation* for the oscillator,

$$
m\ddot{y}(t) = F(y(t)) - 2\dot{y}(t) + 2\dot{w}_{\mathrm{in}}(t), \qquad t > 0; \quad y(0) = \psi_0(0); \quad \dot{y}(0) = p_0, \qquad (1.2.28)
$$

where

$$
w_{\mathrm{in}}(t) = g_+(t) + f_-(-t), \qquad t > 0 \qquad (1.2.29)
$$

is the *incident wave*. Multiplying equation (1.2.28) by $\dot{y}(t)$ and integrating, we get the energy balance

$$\frac{m\dot{y}^2(t)}{2}+U(y(t)) = \frac{m\dot{y}^2(0)}{2}+U(y(0)) -2\int_0^t \dot{y}^2(s)ds +2\int_0^t \dot{w}_{in}(s)\dot{y}(s)ds.$$

$$(1.2.30)$$

Note that

$$\dot{w}_{in} \in L^2(\mathbb{R}^+,\mathbb{R}^d) \qquad (1.2.31)$$

by (1.2.25). Hence (1.2.30) and (1.2.11) imply that the Cauchy problem (1.2.28) admits a unique solution for all $t > 0$, and the a priori bound holds:

$$\sup_{t>0}|y(t)| + \sup_{t>0}|\dot{y}(t)| + \int_0^\infty |\dot{y}(t)|^2 dt \le B < \infty, \qquad (1.2.32)$$

where B is bounded for bounded norm $\|(\psi_0,\pi_0,p_0)\|_{\mathcal{E}}$. These arguments imply that the Cauchy problem (1.2.4) admits a unique solution $Y(t) = (\psi(x,t),\dot{\psi}(x,t),\dot{y}(t)) \in C(\mathbb{R},\mathcal{E})$ for any $Y_0 \in \mathcal{E}$, where $\psi(x,t)$ is defined by (1.2.13), (1.2.24), and (1.2.27) (see [41]).

The a priori bound (1.2.32) implies that $y(t) \in C(\overline{\mathbb{R}^+})$. Hence $y(0)$ exists and

$$f_+(-0) = f_+(+0), \qquad g_-(-0) = g_-(+0) \qquad (1.2.33)$$

since

$$f_+(-0) = y(0) - g_+(+0) = \frac{\psi_0(0)}{2}, \qquad f_+(+0) = \frac{\psi_0(0)}{2} \qquad (1.2.34)$$

and

$$g_-(-0) = \frac{\psi_0(0)}{2}, \qquad g_-(+0) = y(0) - f_-(-0) = \frac{\psi_0(0)}{2} \qquad (1.2.35)$$

by (1.2.26) and (1.2.24).

Corollary 1.2.7 *Formulas (1.2.32) and (1.2.26) imply that*

$$f'_+ \in L^2(\mathbb{R}^-,\mathbb{R}^d), \qquad g'_- \in L^2(\mathbb{R}^+,\mathbb{R}^d) \qquad (1.2.36)$$

by (1.2.25). Hence, (1.2.33) implies that

$$f'_+, g'_- \in L^2(\mathbb{R},\mathbb{R}^d). \qquad (1.2.37)$$

The formulas (1.2.24) and (1.2.27) determine the solution $\psi(x,t)$ uniquely, and $Y(t) := (\psi(x,t),\dot{\psi}(x,t),\psi(0,t)) \in C(\mathbb{R},\mathcal{E})$ due to (1.2.37). Finally, the energy conservation (1.2.15) follows by differentiation; see [41]. Now Proposition 1.2.3 is proved.

Remark 1.2.8 In the energy balance (1.2.30) the integral $2\int_0^t \dot{y}^2(s)ds$ is the energy radiated by the oscillator over the time interval $[0,t]$.

1.2.4 A Relaxation for Reduced Equation

The following lemma on relaxation for the reduced equation plays a crucial role in the proofs of Theorem 1.2.4 and Theorem 1.2.6. Let us denote $\mathcal{Z} = \{(z,0) \in \mathbb{R}^d \times \mathbb{R}^d : z \in Z\}$.

Lemma 1.2.9 *Let all assumptions of Theorem 1.2.4 hold. Then*
(i) For every solution $y(t)$ of the equation (1.2.28),

$$(y(t), \dot{y}(t)) \to \mathcal{Z}, \quad t \to \infty. \tag{1.2.38}$$

(ii) Let, additionally, Z be a discrete subset in $\mathbb{R}^d$. Then there exists a point $(z,0) \in \mathcal{Z}$ such that

$$(y(t), \dot{y}(t)) \to (z,0), \quad t \to \infty.$$

Proof Obviously, (ii) follows from (i). Let us check that (i) follows from (1.2.32). Namely, (1.2.38) is equivalent to the system

$$y(t) \to Z, \quad t \to \infty, \tag{1.2.39}$$
$$\dot{y}(t) \to 0, \quad t \to \infty. \tag{1.2.40}$$

- First, let us prove (1.2.40). Assume the contrary, that

$$|\dot{y}(t_k)| \geq \varepsilon > 0 \tag{1.2.41}$$

for a sequence $t_k \to \infty$. Integrating the equation (1.2.28), we get that

$$m(\dot{y}(t) - \dot{y}(s)) = \int_s^t F(y(\tau))d\tau - 2\int_s^t \dot{y}(\tau)d\tau + 2\int_s^t \dot{w}_{in}(\tau)d\tau, \quad s,t \geq 0. \tag{1.2.42}$$

Let us estimate each of three integrals in the RHS. The first is $\mathcal{O}(|t-s|)$, since $y(\tau)$ is a bounded function by (1.2.32). The second and third integrals are $\mathcal{O}(|t-s|^{1/2})$ by (1.2.32), (1.2.31), and the Cauchy–Schwarz inequality. Hence, (1.2.42) implies that $\dot{y}(t)$ is a Hölder function of degree $1/2$, i.e.,

$$|\dot{y}(t) - \dot{y}(s)| \leq C|t-s|^{1/2}, \quad s,t \geq 0, \quad |t-s| \leq 1.$$

Therefore, $\int_0^\infty \dot{y}^2(t)dt = \infty$ by (1.2.41), which contradicts (1.2.32).
- Now we can prove (1.2.39). Again, assume the contrary. Then

$$F(y(t_k)) \to \overline{F} \neq 0$$

for a sequence $t_k \to \infty$ since $y(t)$ is a bounded function. Moreover, (1.2.40) implies the uniform convergence

$$F(y(\tau)) \to \overline{F}, \quad |\tau - t_k| \leq T$$

for any $T > 0$. Now (1.2.42) and (1.2.40), (1.2.31) imply that

$$m(\dot{y}(t_k + T) - \dot{y}(t_k - T)) = 2T\overline{F} + o(1), \qquad t_k \to \infty,$$

which contradicts (1.2.40) since $\overline{F} \neq 0$.   □

1.2.5 Examples

Let us illustrate Lemma 1.2.9 by an example. For simplicity, let us assume that

$$\psi_0(x) = C_\pm, \quad v_0(x) = 0, \qquad \pm x > r_0$$

with some $C_\pm \in \mathbb{R}$ and $r_0 \geq 0$. Then (1.2.29) implies that $\dot{w}(t) \equiv 0$ for $t > r_0$ and that (1.2.28) is an autonomous equation for $t > r_0$. In the phase plane $(y, \dot{y})$, the orbits of the reduced equation (1.2.28) are determined by the following system:

$$\dot{y}(t) = v(t), \qquad m\dot{v}(t) = F(y(t)) - 2v(t), \qquad t > r_0. \qquad (1.2.43)$$

Let us compare this system with a *free* oscillator that is not coupled to the string,

$$\dot{y} = v, \qquad m\dot{v} = F(y). \qquad (1.2.44)$$

There are simple relationships between phase portraits of these two systems.

A These systems have the same stationary points.

B The vertical component $\dot{v}$ of the phase velocity vector of (1.2.43) is less than that of (1.2.44) if $v > 0$ and is greater if $v < 0$. The horizontal components of these vectors are equal.

C Hence the orbits of (1.2.43) intersect those of (1.2.44) from above in the half-plane $v > 0$ and from below in the half-plane $v < 0$. Let us consider, for instance, a nondegenerate potential of Ginzburg–Landau type

$$V(y) = \frac{1}{4}(y^2 - 1)^2, \quad y \in \mathbb{R}. \qquad (1.2.45)$$

It satisfies conditions (1.2.10) and (1.2.11). Then the system (1.2.44) has the following orbits:

• closed curves corresponding to periodic solutions,
• two separatrices both leaving and entering the point $(0,0)$,
• three stationary points: a saddle at the point $(0,0)$ and two centers at the points $(\pm 1, 0)$ (see Figure 1.2).

Taking into account the property **C**, we see that for the system (1.2.43) with potential (1.2.45),

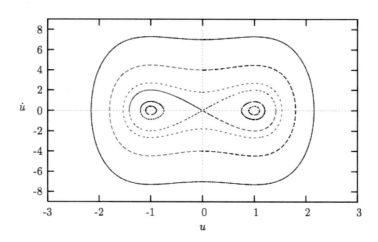

Figure 1.2 Hamiltonian system.

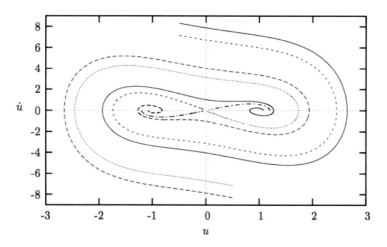

Figure 1.3 System with a friction.

- the points $(\pm 1, 0)$ are stable foci,
- the point $(0, 0)$ is a saddle (see Figure 1.3).

1.2.6 Convergence to Global Attractor

Now we can prove Theorem 1.2.4. It suffices to prove it for $t \to \infty$.

Lemma 1.2.10 *Let all the assumptions of Theorem 1.2.4 hold. Then*

$$Y(t) \xrightarrow{\mathcal{E}_F} \mathcal{S} \text{ as } t \to \infty.$$

Proof It suffices to construct $z(t) \in Z$ for $t \geq 0$ such that

$$\|Y(t) - S_{z(t)}\|_R \to 0 \text{ as } t \to \infty.$$

The convergence (1.2.39) means that there exists a function $z(t) \in Z, t \geq 0$, such that

$$|y(t) - z(t)| \to 0, \qquad t \to \infty. \tag{1.2.46}$$

By definitions (1.2.8) and (1.2.16),

$$\|Y(t) - S_{z(t)}\|_R = \|\psi'(\cdot,t)\|_R + |\psi(0,t) - z(t)| + \|\dot\psi(\cdot,t)\|_R + |\dot y(t)|.$$

Here both norms $\|\ldots\|_R \to 0$ due to (1.2.13), (1.2.25), (1.2.36) and (1.2.37). Therefore, (1.2.46) and (1.2.40) complete the proof. □

Now Theorem 1.2.4 (i) is proved. Then Theorem 1.2.4 (ii) follows, since the set $\mathcal{S}$, isomorphic to Z, is discrete.

Remark 1.2.11 The bound (1.2.32) is provided by the friction term in the reduced equation (1.2.28) for the nonlinear oscillator. The friction means the energy radiation by the oscillator, and the integral in (1.2.32) represents the energy radiated to infinity. Thus, our proof of Theorem 1.2.4 relies on the energy radiation to infinity.

1.2.7 The Transitivity of the Transitions

The next lemma shows that the transitions of type (8) exist for any two stationary states $S_\pm$.

Lemma 1.2.12 *Let conditions of Theorem 1.2.4 hold. Then, for every two stationary states $S_\pm \in \mathcal{S}$, there exist solutions $Y(t) \in C(\mathbb{R}, \mathcal{E})$ to the system (1.7.13), intertwining $S_\pm$ in the sense (1.2.19).*

Proof Let $S_\pm = (s_\pm(x), 0, 0)$ with $s_\pm(x) \equiv z_\pm \in Z$. It is possible to provide the transition $S_- \mapsto S_+$ in different ways. We choose one of them, which is possibly most obvious. Namely, we construct a solution $Y(t) = (u(\cdot,t), \dot u(\cdot,t), y(t)) \in C(\mathbb{R}, \mathcal{E})$ to (1.7.13) such that

$$y(t) := u(0,t) = \begin{cases} z_- & \text{for } t \leq -1, \\ z_+ & \text{for } t \geq 1. \end{cases} \tag{1.2.47}$$

We extend $y(t)$ for $t \in (-1,1)$ arbitrarily so that $y \in C^2(\mathbb{R}, \mathbb{R}^d)$. Then we set $g_+(z) \equiv z_-$ and determine f_- by (1.2.28):

$$m\ddot y(t) = F(y(t)) + 2(f'_-(-t) - \dot y(t)), \quad t \in \mathbb{R}. \tag{1.2.48}$$

Then $f'_-(z) \in C(\mathbb{R}, \mathbb{R}^d)$. Since $F(z_\pm) = 0$, we have

$$f'_-(-t) \equiv 0 \quad \text{for} \quad t \le -1 \quad \text{and for} \quad t \ge 1. \tag{1.2.49}$$

To determine f_- uniquely, we may require that

$$f_-(-t) \equiv z_- \quad \text{for} \quad t \le -1. \tag{1.2.50}$$

Then the reflected waves g_- and f_+ are determined by (1.2.26).

Since $y(t)$, $f_-(-t)$, and $g_+(t)$ are constant for large $|t|$, $f_+(-t)$, $g_-(t)$ are also constant for large $|t|$. Then, for $u(x,t)$ defined by (1.2.27), the function

$$Y(t) = (u(\cdot,t), \dot{u}(\cdot,t), \dot{u}(0,t)) \in C(\mathbb{R}, \mathcal{E})$$

is a solution to (1.2.1), and (1.2.19) holds. $\square$

Remark 1.2.13 Physically, the inequality $z_+ \ne z_-$ means the capture of radiation by the oscillator if $V(z_+) > V(z_-)$ or the emission of radiation by the oscillator if $V(z_+) < V(z_-)$.

1.2.8 Divergent Wave

Here we prove Theorem 1.2.6. First, let us construct the divergent wave

$$\tilde{W}(t)\Psi_+ = (\mathrm{w}_{\mathrm{out}}(x,t), \dot{\mathrm{w}}_{\mathrm{out}}(x,t), 0), \quad t \ge 0.$$

Here $\mathrm{w}_{\mathrm{out}}(x,t)$ is a finite-energy solution to the free d'Alembert equation. Let us set

$$\mathrm{w}_{\mathrm{out}}(x,t) = C_0 + f_+(x-t) + g_-(x+t), \tag{1.2.51}$$

where the constant C_0 will be chosen below. It remains to check (1.2.22) and (1.2.23) for $t \to \infty$, which means the representation

$$(\psi(x,t), \dot{\psi}(x,t), \dot{y}(t)) = (s_+(x), 0, 0)$$
$$+ (\mathrm{w}_{\mathrm{out}}(x,t), \dot{\mathrm{w}}_{\mathrm{out}}(x,t), 0) + r_+(t), \quad t > 0,$$

where

$$s_+(x) \equiv z_+ := \lim_{t \to +\infty} y(t) \tag{1.2.52}$$

and

$$\|r_+(t)\|_{\mathcal{E}} \to 0, \quad t \to +\infty. \tag{1.2.53}$$

By definition of the norm (1.2.7), (1.2.53) is equivalent to

$$\|\psi'(\cdot,t) - w'_{out}(\cdot,t)\|_{L^2(\mathbb{R},\mathbb{R}^d)} + |\psi(0,t) - z_+ - w_{out}(0,t)|$$

$$+ \|\dot\psi(\cdot,t) - \dot w_{out}(\cdot,t)\|_{L^2(\mathbb{R},\mathbb{R}^d)} \to 0, \quad t \to \infty \tag{1.2.54}$$

since $\dot y(t) \to 0$ by (1.2.40).

Step (i). Let us start with the second term in the LHS of (1.2.54). Since $\psi(0,t) = y(t) \to z_+$, it suffices to prove that

$$w_{out}(0,t) = C_0 + f_+(-t) + g_-(t) \to 0, \quad t \to +\infty. \tag{1.2.55}$$

First, (1.2.6) and (1.2.24) imply that

$$\lim_{t\to\infty} f_-(-t) = \frac{\psi_0^-}{2} - \frac{1}{2}\int_0^{-\infty} \pi_0(y)dy,$$

$$\lim_{t\to+\infty} g_+(t) = \frac{\psi_0^+}{2} + \frac{1}{2}\int_0^{\infty} \pi_0(y)dy. \tag{1.2.56}$$

Second, we have by (1.2.26) and (1.2.52) that

$$\lim_{t\to\infty} f_+(-t) = z_+ - \lim_{t\to+\infty} g_+(t); \quad \lim_{t\to+\infty} g_-(t) = z_+ - \lim_{t\to\infty} f_-(-t).$$

Substituting (1.2.56), we obtain

$$\begin{cases} \displaystyle\lim_{t\to\infty} f_+(-t) = z_+ - \frac{\psi_0^+}{2} - \frac{1}{2}\int_0^{\infty} \pi_0(y)dy, \\[2mm] \displaystyle\lim_{t\to+\infty} g_-(t) = z_+ - \frac{\psi_0^-}{2} + \frac{1}{2}\int_0^{-\infty} \pi_0(y)dy. \end{cases}$$

Hence, (1.2.55) holds if we choose

$$C_0 := \frac{\psi_0^+}{2} + \frac{\psi_0^-}{2} + \frac{I_0}{2} - 2z_+, \tag{1.2.57}$$

where I_0 is defined in (1.2.6).

Step (ii). Now, let us consider the first term in the LHS of (1.2.54). It suffices to prove, for example, that

$$\|\psi'(\cdot,t) - w'_{out}(\cdot,t)\|_{L^2(\mathbb{R}^+,\mathbb{R}^d)} \to 0, \quad t \to \infty.$$

Using (1.2.51) and the d'Alembert representation (1.2.13) for $x > 0$, we get

$$\psi'(x,t) - w'_{out}(x,t) = g'_+(x+t) - g'_-(x+t), \quad x \geq t.$$

Finally, (1.2.25) and (1.2.36) imply that

$$\|g'_+(x+t) - g'_-(x+t)\|^2_{L^2(\mathbb{R}^+, \mathbb{R}^d)}$$

$$\leq C \int_0^\infty \left[|g'_+(x+t)|^2 + |g'_-(x+t)|^2 \right] dx$$

$$= C \int_t^\infty \left[|g'_+(z)|^2 + |g'_-(z)|^2 \right] dz \to 0, \quad t \to \infty.$$

Step (iii). The third term in the LHS of (1.2.54) can be handled similarly. Theorem 1.2.6 is proved.

1.3 String Coupled to Several Nonlinear Oscillators

Here we present the results [42], which extend the results of the previous section 1.2 to the case of a string with several nonlinear oscillators:

$$\ddot{\psi}(x,t) = \psi''(x,t) + \sum_1^N \delta(x - x_k) F_k(\psi(x_k, t)), \qquad x \in \mathbb{R}.$$

This equation reduces to a system of N ordinary differential equations with delay. Its study required a new approach relying on a special analysis of a *relaxation* of all trajectories.

1.3.1 Introduction

Let $Q = \{x_1, \ldots, x_N\}$ be a finite set of N points $x_k \in \mathbb{R}$. We establish global attraction to stationary states (7) for all finite-energy solutions to the system of equations

$$\ddot{\psi}(x,t) = \psi''(x,t), \quad x \in \mathbb{R} \setminus Q, \tag{1.3.1}$$

together with the gluing conditions at the points $x_k \in Q$,

$$\left\{ \begin{array}{l} \psi(x_k + 0, t) = \psi(x_k - 0, t) \\ \\ 0 = F_k(\psi(x_k, t)) + \psi'(x_k + 0, t) - \psi'(x_k - 0, t) \end{array} \right. \Bigg|. \tag{1.3.2}$$

In the case $N = 1$, this system coincides with the Lamb system (1.2.1) with $m = 0$. The solutions $\psi(x,t)$ take the values in $\mathbb{R}^d$ with $d \geq 1$. Note that the system (1.3.1) is formally equivalent to the 1D nonlinear wave equation with the nonlinear term concentrated at the set Q (cf. (1.2.2)),

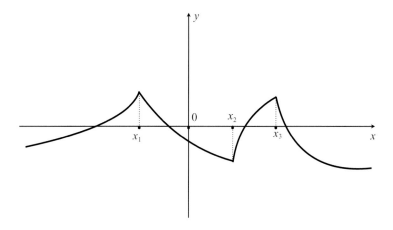

Figure 1.4 String coupled to nonlinear oscillators.

$$\ddot{\psi}(x,t) = \psi''(x,t) + \sum_{k=1}^{N} \delta(x - x_k) F_k(\psi(x_k,t)), \qquad x \in \mathbb{R}. \qquad (1.3.3)$$

Physically, the system (1.3.1), (1.3.2) describes small crosswise oscillations of a string that is subject to constraint forces F_k at the points x_k, and the forces are perpendicular to the string. For example, $F_k(y) = -\omega_k^2 y$ if the string is attached to a linear spring at the point x_k (see Figure 1.4). But in general, the functions $F_k(y)$ are nonlinear.

We introduce the Hilbert phase space $\mathcal{E}$ of finite-energy states for the system (1.3.1), (1.3.2).

Definition 1.3.1 (i) $\mathcal{E} = E_c \oplus L^2$ is the Hilbert space of pairs $(\psi(x), \pi(x))$, with the norm

$$\|(\psi,\pi)\|_{\mathcal{E}} = \|\psi\|_{E_c} + \|\pi\|. \qquad (1.3.4)$$

(ii) $\mathcal{E}_F$ is the space $\mathcal{E}$ endowed with the topology defined by the seminorms

$$\|(\psi,\pi)\|_R \equiv \|\psi'\|_R + |\psi(0)| + \|\pi\|_R, \quad R > 0. \qquad (1.3.5)$$

We assume the following conditions:

$$\begin{cases} \text{all } F_k \in C^1(\mathbb{R}^d, \mathbb{R}^d), \ F_k(\psi) = -\nabla V_k(\psi) \\ \inf_{y \in \mathbb{R}^d} V_k(y) > -\infty, \qquad \forall k = 1, \dots, N \\ V_k(y) \to +\infty \ \text{ as } |y| \to \infty \ \text{ for some } k = 1, \dots, N \end{cases}. \qquad (1.3.6)$$

Then the system (1.3.1), (1.3.2) is formally Hamiltonian with the Hilbert phase space $\mathcal{E}$ and the Hamiltonian functional

$$\mathcal{H}(\psi,\pi) = \frac{1}{2}\int_{\mathbb{R}}[|\pi(x)|^2 + |\psi'(x)|^2]dx + \sum_{k=1}^{N} V_k(\psi(x_k)), \qquad (\psi,\pi)\in\mathcal{E}.$$

$$(1.3.7)$$

We consider solutions $Y(t) = (\psi(\cdot,t),\dot\psi(\cdot,t)) \in C(\mathbb{R},\mathcal{E})$, and we write the system (1.3.1), (1.3.2) in the form

$$\dot Y(t) = \mathbf{F}(Y(t)), \qquad t\in\mathbb{R}. \qquad (1.3.8)$$

Let us discuss the definition of the Cauchy problem for the functions $Y(t)\in C(\mathbb{R},\mathcal{E})$. The first equation of (1.3.2) makes sense and holds automatically because $\psi\in C(\mathbb{R}^2,\mathbb{R}^d)$ by the Sobolev embedding theorem due to $Y(t)\in C(\mathbb{R},\mathcal{E})$. The equation (1.3.1) is understood in the sense of distributions of $(x,t)\in[\mathbb{R}\setminus Q]\times\mathbb{R}$. Hence, this equation is equivalent to the d'Alembert decompositions for every $k=1,\dots,N+1$,

$$\psi(x,t) = f_k(x-t) + g_k(x+t), \qquad x\in\Delta_k := (x_{k-1},x_k),\ t\in\mathbb{R}, \quad (1.3.9)$$

where $f_k,g_k\in C(\mathbb{R},\mathbb{R}^d)$ due to $\psi\in C(\mathbb{R}^2,\mathbb{R}^d)$, and we denote $x_0 := -\infty$ and $x_{N+1} = +\infty$. Hence, for all $k=1,\dots,N$ and $(x,t)\in\Delta_k\times\mathbb{R}$,

$$\psi'(x,t) = f_k'(x-t) + g_k'(x+t), \quad \dot\psi(x,t) = -f_k'(x-t) + g_k'(x+t),$$

$$(1.3.10)$$

where all derivatives are understood in the sense of distributions. The assumption $Y(t)\in C(\mathbb{R},\mathcal{E})$ implies

$$f_k'(\cdot), g_k'(\cdot)\in L^2_{\mathrm{loc}}(\mathbb{R},\mathbb{R}^d), \quad \forall k=1,\dots,N+1. \qquad (1.3.11)$$

We now explain the second equation of (1.3.2).

Definition 1.3.2 In the second equation of (1.3.2), for every $k=1,\dots,N$,

$$\left\{\begin{array}{l} \psi'(x_k-0,t) := f_k'(x_k-t) + g_k'(x_k+t) \in L^2_{\mathrm{loc}}(\mathbb{R},\mathbb{R}^d) \\ \psi'(x_k+0,t) := f_{k+1}'(x_k-t) + g_{k+1}'(x_k+t) \in L^2_{\mathrm{loc}}(\mathbb{R},\mathbb{R}^d) \end{array}\right. . \quad (1.3.12)$$

Note that the functions f_k and g_k in (1.3.9) are unique up to an additive constant, so the definition (1.3.12) is unambiguous.

1.3.2 Main Results

We start with the existence of the dynamics.

Proposition 1.3.3 *Let $d \geq 1$ and assumptions (1.3.6) hold. Then*

(i) For every initial state $Y(0) \in \mathcal{E}$, equation (1.3.8) has a unique solution $Y(t) \in C(\mathbb{R}, \mathcal{E})$.

(ii) The mapping $W(t) : Y(0) \mapsto Y(t)$ is continuous in $\mathcal{E}$ and in $\mathcal{E}_F$ for every $t \in \mathbb{R}$.

(iii) The energy (1.3.7) is conserved,

$$\mathcal{H}(Y(t)) = \text{const}, \qquad t \in \mathbb{R}. \tag{1.3.13}$$

This proposition will be proved in the next section.

Definition 1.3.4 $\mathcal{S}$ denotes the set of all stationary states $S = (s(x), 0) \in \mathcal{E}$ of the system (1.3.8).

The next proposition gives a criterion for the set $\mathcal{S}$ being a nonempty discrete subset of $\mathcal{E}_F$.

Proposition 1.3.5 *Let conditions (1.3.6) hold, $d = 1$, and all functions $F_k(y)$ with $k = 1, \ldots, N$ be real-analytic on $\mathbb{R}$. Then $\mathcal{S}$ is a discrete subset of $\mathcal{E}_F$.*

The main result of this section means that the set $\mathcal{S}$ is the global attractor of the system (1.3.8) in the topology of the space $\mathcal{E}_F$.

Theorem 1.3.6 *Let $d \geq 1$, assumptions (1.3.6) hold, and an initial state $Y(0) \in \mathcal{E}$. Then*

   (i) The corresponding solution $Y(t) \in C(\mathbb{R}, \mathcal{E})$ of equation (1.3.8), attracts to the set $\mathcal{S}$ in the sense (1.2.18)

$$Y(t) \xrightarrow{\mathcal{E}_F} \mathcal{S}, \qquad t \to \pm\infty. \tag{1.3.14}$$

(ii) Let, moreover, $d = 1$ and all functions $F_k(y_k)$ be real-analytic on $\mathbb{R}$. Then any solution $Y(t) \in C(\mathbb{R}, \mathcal{E})$ attracts to some stationary states $S_\pm \in \mathcal{S}$ depending on the solution,

$$Y(t) \xrightarrow{\mathcal{E}_F} S_\pm, \qquad t \to \pm\infty. \tag{1.3.15}$$

Remarks 1.3.7 (i) The assertion (ii) of this theorem follows from (i) due to Proposition 1.3.5.

(ii) The convergence (1.3.15) and (1.3.7), (1.3.6) imply (10) by the Fatou theorem.

1.3.3 Well-Posedness and A Priori Estimates

Proof of Proposition 1.3.3 The solution $Y(t) \in C(\mathbb{R}, \mathcal{E})$ to (1.3.8) can be constructed by the d'Alembert representations (1.3.9) similarly to the case $N = 1$, considered in Section 1.2. However, for $N > 1$, we need to find repeatedly reflected waves from all points x_k with $k = 1, \ldots, N$. The energy conservation (1.3.13) follows by methods of [41] using the d'Alembert representations (1.3.9). $\qquad\square$

Let us show that the energy conservation implies the following a priori estimate, which we will need in the proof of Theorem 1.3.6.

Proposition 1.3.8 *Let the conditions (1.3.6) hold. Then, for every solution $Y(t) \in C(\mathbb{R}, \mathcal{E})$ of (1.3.8), all functions $y_k(t) := \psi(x_k, t)$ are bounded:*

$$\sup_{t \in \mathbb{R}} |y_k(t)| < \infty, \qquad k = 1, \ldots, N. \tag{1.3.16}$$

Proof We prove in fact a slightly stronger statement. Namely, denote $y_k = y_k(\psi) = \psi(x_k)$ and $\overline{y} = \overline{y}(\psi) = (y_1, \ldots, y_N)$ for $\psi \in E_c$. Denote by $\mathcal{U}$ the potential energy functional:

$$\mathcal{U}(\psi) \equiv \mathcal{H}(\psi, 0) = \frac{1}{2} \int_{-\infty}^{\infty} |\psi'(x)|^2 \, dx + \sum_{k=1}^{N} V_k(y_k), \qquad \psi \in E_c. \tag{1.3.17}$$

Then (1.3.16) follows from

$$\mathcal{U}(\psi) \to \infty \quad as \quad |\overline{y}(\psi)| \to \infty. \tag{1.3.18}$$

To prove this, it suffices to show that

$$\sup_{\mathcal{U}(\psi) \leq E} |\overline{y}(\psi)| < \infty \tag{1.3.19}$$

for every $E \in \mathbb{R}$. First, all potentials V_k are bounded below by (1.3.6). Hence,

$$\sup_{\mathcal{U}(\psi) \leq E} \int_{-\infty}^{\infty} |\psi'(x)|^2 dx = D < \infty. \tag{1.3.20}$$

Second, the Cauchy–Schwarz inequality gives, for every $k, j = 1, \ldots, N$,

$$\sup_{\mathcal{U}(\psi) \leq E} |y_k - y_j| = \sup_{\mathcal{U}(\psi) \leq E} \left| \int_{x_k}^{x_j} \psi'(x) \, dx \right| \leq |x_k - x_j|^{1/2} D^{1/2}. \tag{1.3.21}$$

Therefore, (1.3.19) follows from the last condition of (1.3.6). $\qquad\square$

1.3.4 Stationary States

In this section we prove Proposition 1.3.5. Substituting $\psi(x,t) = s(x)$ for (1.3.1), we obtain that $s''(x) = 0$ for $x \in \mathbb{R} \setminus Q$. Hence,

$$s(x) = a_k x + b_k \quad \text{for} \quad x \in \Delta_k := (x_{k-1}, x_k), \quad k = 1, \ldots, N + 1,$$
$$(1.3.22)$$

where $x_0 := -\infty$ and $x_{N+1} := +\infty$. The condition $s' \in L^2(\mathbb{R})$ implies

$$a_1 = a_{N+1} = 0. \tag{1.3.23}$$

Substituting (1.3.22) into equations (1.3.2), we obtain that

$$\begin{cases} a_k x_k + b_k = y_k = a_{k+1} x_k + b_{k+1} \\ 0 = F_k(y_k) + a_{k+1} - a_k \end{cases}, \quad k = 1, \ldots, N. \tag{1.3.24}$$

Hence, equations (1.3.23) imply that the function (1.3.22) is uniquely defined by its values $y_k = s(x_k)$ at the points $x_k, k = 1, \ldots, N$:

$$a_k = \frac{y_k - y_{k-1}}{l_k}, \quad b_k = y_k - a_k x_k, \quad k = 1, \ldots, N + 1. \tag{1.3.25}$$

Here $y_0 := y_1$, $y_{N+1} := y_N$, $l_k := x_k - x_{k-1}$ for $k = 2, \ldots, N$, and $l_1 := 1$, $l_{N+1} := 1$ (for instance). For unknown $y_k, k = 1, \ldots, N$, the system (1.3.24) is equivalent to

$$F_k(y_k) + \frac{y_{k+1} - y_k}{l_{k+1}} - \frac{y_k - y_{k-1}}{l_k} = 0, \quad k = 1, \ldots, N. \tag{1.3.26}$$

Since $s(x) \in C^2(\overline{\Delta_k})$, the variation $DU(s)$ exists and

$$-DU(s) = s''(x) + \sum_{k=1}^{N} (s'(x_k + 0) - s'(x_k - 0) - \nabla V_k(y_k)) \delta(x - x_k).$$

Therefore, the system (1.3.1), (1.3.2) for stationary states implies the variation equation

$$DU(s) = 0. \tag{1.3.27}$$

Proof of Proposition 1.3.5 Let us define the function in $\mathbb{R}^N$

$$U_N(y_1, \ldots, y_N) = U(s), \tag{1.3.28}$$

where $s = s(x)$ is the stationary solution (1.3.22) with a_k and b_k defined by (1.3.25). Then (1.3.17) implies

$$\mathcal{U}_N(y_1, \ldots, y_N) = \frac{1}{2} \sum_{k=2}^{N} \left| \frac{y_k - y_{k-1}}{l_k} \right|^2 l_k + \sum_{k=1}^{N} V_k(y_k). \qquad (1.3.29)$$

Now (1.3.27) gives for stationary solutions

$$\frac{\partial \mathcal{U}_N}{\partial y_k}(y_1, \ldots, y_N) = 0, \quad k = 1, \ldots, N. \qquad (1.3.30)$$

On the other hand, (1.3.18) implies

$$\mathcal{U}_N(y_1, \ldots, y_N) \to \infty \quad \text{as} \quad |(y_1, \ldots, y_N)| \to \infty. \qquad (1.3.31)$$

Hence, $\mathcal{U}_N$ gets a minimal value at a certain point $(y_1, \ldots, y_N) \in R^N$, so $\mathcal{S} \neq \emptyset$.

Take $y_0(\lambda) = y_1(\lambda) = \lambda \in \mathbb{R}$. Then we can define uniquely $y_2(\lambda), \ldots, y_N(\lambda)$ in a sequel according to formulas (1.3.26) with $k = 1, \ldots, N - 1$. Therefore, the continuous map $I_1 : \mathcal{E}_F \to \mathbb{R}^d$ defined by

$$I_1(\psi(x), \pi(x)) = \psi(x_1)$$

is an isomorphism on $\mathcal{S}$. Hence, Proposition 1.3.5 obviously follows from the next lemma.

Lemma 1.3.9 $Z_1 := I_1 \mathcal{S}$ *is a discrete subset of* $\mathbb{R}$.

Proof All functions $y_k(\lambda)$ are real-analytic on $\mathbb{R}$ for $k = 2, \ldots, N$. The last equation of (1.3.24) with $k = N$ gives

$$a_{N+1} = a_N - F_N(y_N) = \frac{y_N - y_{N-1}}{l_N} - F_N(y_N). \qquad (1.3.32)$$

The vector $\{y_k(\lambda) : k = 1, \ldots, N\}$ defines the stationary solution $s_\lambda(x)$ via (1.3.25), (1.3.22) if and only if $a_{N+1} = 0$. Thus, we get the following equation for $\lambda \in Z_1$:

$$T(\lambda) := \frac{y_N(\lambda) - y_{N-1}(\lambda)}{l_N} - F_N(y_N(\lambda)) = 0. \qquad (1.3.33)$$

The map $\lambda \mapsto T(\lambda)$ is real-analytic on $\lambda \in \mathbb{R}$. Hence, the set Z_1 of all solutions to (1.3.33) is either a discrete set in $\mathbb{R}$ or $Z_1 = \mathbb{R}$.

Let us show that the case $Z_1 = \mathbb{R}$ is impossible under conditions (1.3.6) even if the functions F_k are not real-analytic. Assume the converse: $Z_1 = \mathbb{R}$. Then

$$\mathcal{U}_N(y_1(\lambda), \ldots, y_N(\lambda)) = \text{const}, \qquad \lambda \in \mathbb{R}. \qquad (1.3.34)$$

Indeed, since $F_k \in C^1(\mathbb{R})$, we have $y_k(\lambda) \in C^1(\mathbb{R})$ for all $k = 1, \ldots, N$. Then, by (1.3.30), we obtain

$$\partial_\lambda \mathcal{U}_N(y_1(\lambda), \ldots, y_N(\lambda)) = \sum_{k=1}^{N} \frac{\partial \mathcal{U}_N}{\partial y_k} y_k'(\lambda) = 0, \quad \lambda \in \mathbb{R}. \qquad (1.3.35)$$

On the other hand, (1.3.29) implies

$$\mathcal{U}_N(y_1(\lambda), \ldots, y_N(\lambda)) = \frac{1}{2} \sum_{k=2}^{N} \left| \frac{y_k(\lambda) - y_{k-1}(\lambda)}{l_k} \right|^2 l_k + \sum_{k=1}^{N} V_k(y_k(\lambda)).$$
$$(1.3.36)$$

Therefore, (1.3.34) and the middle condition (1.3.6) imply that the first sum on the RHS of (1.3.36) is bounded for $\lambda \in \mathbb{R}$. Hence,

$$y_k(\lambda) \to \infty \quad \text{as} \quad |y_1(\lambda)| = |\lambda| \to \infty, \quad \forall k = 2, \ldots, N. \qquad (1.3.37)$$

However, then the second sum on the RHS of (1.3.29) tends to infinity as $|\lambda| \to \infty$ due to the last condition of (1.3.6). Hence,

$$\mathcal{U}_N(y_1(\lambda), \ldots, y_N(\lambda)) \to \infty \quad \text{as} \quad |\lambda| \to \infty, \qquad (1.3.38)$$

which contradicts (1.3.34). $\qquad\qquad\qquad\qquad\qquad\qquad\qquad\qquad\qquad$ □

1.3.5 Examples

In this section we consider examples of systems (1.3.1) with $d = 1$.

Example 1.3.10 Let each potential $V_k(y)$ be a polynomial of an even degree $p_k + 1 \geq 2$ with positive leading coefficient. Then all functions $F_k(y) = -\nabla V_k(y)$ are polynomials of degrees $p_k \geq 1$ and all conditions of Proposition 1.3.5 hold. By (1.3.26), each function $y_k(\lambda)$, $k \geq 2$ is a polynomial of degree less than or equal to the product $p_1 \ldots p_{k-1}$. Hence, the equation (1.3.33) has no more than $\overline{p} := p_1 \ldots p_N$ roots $\lambda \in \mathbb{R}$, and the set $\mathcal{S}$ has no more than $\overline{p}$ points.

The next examples show that if the potentials V_k do not satisfy either some of conditions (1.3.6) or the analyticity condition, then the set $\mathcal{S}$ can be nondiscrete.

Example 1.3.11 The middle and the last conditions of (1.3.6) break down for the system (1.3.1) with $N = 2$, $x_1 = -1$, $x_2 = 1$, and

$$V_k(y) = -\frac{y^2}{2}, \quad k = 1, 2. \qquad (1.3.39)$$

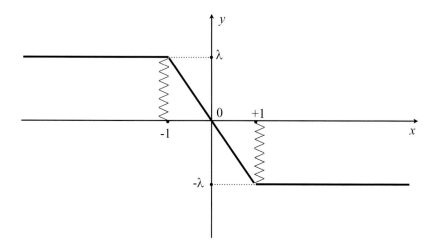

Figure 1.5 Stationary states.

Then $F_k(y) = y$ is the force repulsing from the equilibrium position $y = 0$. In this case the system (1.3.1) has a continuum of solutions of the type (see Figure 1.5)

$$s_\lambda(x) = \begin{cases} \lambda, & x \le -1, \\ -\lambda x, & -1 \le x \le 1, \\ -\lambda, & x \ge 1. \end{cases} \quad (1.3.40)$$

Here $y_1 = s_\lambda(-1) = \lambda$ is an arbitrary real number, so $Y_1 = \mathbb{R}$. The potentials $V_k(y)$ are real-analytic.

The last condition of (1.3.6) can be formally provided by introduction of the elastic force $F_3(y) = -y$ with the potential $V_3(y) = y^2/2$ at the point $x_3 = 0$. Then the functions (1.3.40) remain stationary solutions to the new system involving the three forces, since $s_\lambda(0) = 0$ for all $\lambda \in \mathbb{R}$. So the first and last conditions of (1.3.6) and the analyticity condition hold, but the middle condition of (1.3.6) breaks down, and the set $\mathcal{S}$ is not discrete.

Example 1.3.12 The last condition (1.3.6) breaks down for the system with $V_k(y) \equiv C_k$ for all k. In this case,

$$F_k(y) \equiv 0, \qquad y \in \mathbb{R}.$$

Then $s_\lambda(x) \equiv \lambda$ for $x \in \mathbb{R}$ is the stationary solution to the system (1.3.1) for any $\lambda \in \mathbb{R}$. Thus, $Y_1 = \mathbb{R}$, as in the previous example. The first and the middle conditions of (1.3.6) and the analyticity condition hold, but the last condition of (1.3.6) breaks down, and the set $\mathcal{S}$ is not discrete.

Example 1.3.13 Now let us neglect the analyticity condition. Consider potentials $V_k(y)$ such that

(i) $V_k(y) \in C^2(\mathbb{R})$ satisfy all conditions (1.3.6).

(ii) $V_k(y) \to \infty$ as $|y| \to \infty$ for every $k = 1, \dots, N$.

(iii) $V_k(y) \equiv C_k$ for $y \in [a, b]$, where $a < b$. Then

$$F_k(y) \equiv 0, \qquad y \in [a, b], \quad \forall k = 1, \dots, N. \qquad (1.3.41)$$

It is clear that such functions V_k exist and are not analytic. Hence, the functions $s_\lambda(x) \equiv \lambda$ are stationary solutions to the system (1.3.1), if $\lambda \in [a, b]$. Thus, the set $\mathcal{S}$ is not discrete, though all conditions (1.3.6) hold. Let us note, however, that $Z_1 \neq \mathbb{R}$ here in accordance with Lemma 1.3.9.

Remark 1.3.14 In Examples 1.3.12 and 1.3.13 the global attraction (1.3.14) holds while (1.3.15) breaks down. Namely, each function $f(x - t)$ with values in the interval $[a, b]$ is a solution to the system (1.3.1). It is easy to construct such solution with $(\psi, \dot\psi) \in C(\mathbb{R}, \mathcal{E})$. For example, in the case $a = -1$ and $b = 1$, we can take the function (1.2.21).

1.3.6 Long-Time Asymptotics

In this section we prove Theorem 1.3.6.

Compact Attracting Set and Global Attraction

First, we construct a finite-dimensional attracting set $\mathcal{A}$. The set consists of piecewise linear functions (1.3.22). Namely, for any $\alpha = \{(a_k, b_k) \in \mathbb{R}^{2d} : k = 1, \dots, N + 1\} \in (\mathbb{R}^{2d})^{N+1}$, let us denote

$$\psi_\alpha(x) = a_k x + b_k, \qquad x \in \Delta_k, \quad k = 1 \dots, N + 1 \qquad (1.3.42)$$

and

$$A_{\mathcal{E}} = \{\alpha \in (\mathbb{R}^{2d})^{N+1} : \psi_\alpha(x_k - 0)$$
$$= \psi_\alpha(x_k + 0), \ k = 1, \dots, N; a_1 = a_{N+1} = 0\}.$$

Then $(\psi_\alpha(x), 0) \in \mathcal{E}$ for every $\alpha \in A_{\mathcal{E}}$.

Definition 1.3.15 $\mathcal{A} = \{S_\alpha = (\psi_\alpha(x), 0) : \alpha \in A_{\mathcal{E}}\}$.

Obviously, $\mathcal{A}$ is a locally compact subset in $\mathcal{E}_F$. We prove the next lemma in the following section.

Lemma 1.3.16 *Let all assumptions of Theorem 1.3.6 hold. Then*

$$Y(t) \xrightarrow{\mathcal{E}_F} \mathcal{A}, \qquad t \to \infty. \tag{1.3.43}$$

Let us deduce Theorem 1.3.6 from this lemma.

Definition 1.3.17 Denote by $\omega(Y)$ the omega-set of the trajectory $Y(t)$ in the topology of the space $\mathcal{E}_F$: $\overline{Y} \in \omega(Y)$ if and only if

$$Y(t_k) \xrightarrow{\mathcal{E}_F} \overline{Y} \tag{1.3.44}$$

for some sequence $t_k \to \infty$.

The following lemma implies (1.3.14).

Lemma 1.3.18 *(i) $\omega(Y) \neq \emptyset$ and (ii) $\omega(Y) \subset \mathcal{S}$.*

Proof (i) Lemma 1.3.16 means that there exists a function $\alpha(t) \in C[0,\infty; A_{\mathcal{E}})$ such that for every $R > 0$,

$$\|Y(t) - S_{\alpha(t)}\|_R \to 0 \quad \text{as} \quad t \to +\infty. \tag{1.3.45}$$

Here $S_{\alpha(t)} = (\psi_{\alpha(t)}, 0)$ and $\psi_{\alpha(t)}(x)$ is defined by (1.3.42) with $\alpha = \alpha(t) = \{(a_k(t), b_k(t)) \in \mathbb{R}^{2d} : k = 1, \dots, N+1\}$.

The orbit $\{S_{\alpha(t)} : t > 0\}$ is precompact in $\mathcal{E}_F$ by the bounds (1.3.16). Hence, the limit (1.3.45) implies that the orbit $\{Y(t) : t > 0\}$ also is precompact in $\mathcal{E}_F$. Therefore, $\omega(Y) \neq \emptyset$.

(ii) $\omega(Y) \subset \mathcal{A}$ by (1.3.43). Moreover, the set $\omega(Y)$ is invariant with respect to dynamical group $W(t)$ due to the continuity of $W(t)$ in $\mathcal{E}_F$. Hence, for every $\overline{Y} \in \omega(Y)$, there exists a C^1-curve $t \mapsto \alpha(t) \in A_{\mathcal{E}}$ such that $W(t)\overline{Y} = S_{\alpha(t)}$. Then $S_{\alpha(t)}(x) = (\psi_{\alpha(t)}(x), 0)$ is a solution to the system (1.3.8). In particular, $\partial_t \psi_{\alpha(t)}(x) \equiv 0$. Therefore, $\alpha(t) \equiv \alpha$ and $\overline{Y} = S_\alpha \in \mathcal{S}$. $\qquad\square$

1.3.7 Attraction to a Compact Set

It remains to prove Lemma 1.3.16. It suffices to construct a function $\alpha(t) \in C[0,\infty; A_{\mathcal{E}})$ satisfying (1.3.45).

We may assume without loss of generality that $x_1 = 0$. Then $\psi_{\alpha(t)}(0) = b_1(t)$, and (1.3.45) according to the definition of the norm (1.3.5) means that

$$\int_{-R}^{R} |\psi'(x,t) - \psi'_{\alpha(t)}(x)|^2 \, dx$$

$$+ \int_{-R}^{R} |\dot{\psi}(x,t)|^2 \, dx + |\psi(0,t) - b_1(t)| \to 0, \quad t \to \infty. \tag{1.3.46}$$

We choose $b_1(t) = y_1(t)$ for $t > 0$. Then (1.3.46) for $R > \max(|x_1|, |x_N|)$ becomes

$$\int_{-R}^{x_1} |\psi'(x,t)|^2\, dx + \sum_{2 \le k \le N} \int_{x_{k-1}}^{x_k} |\psi'(x,t) - a_k(t)|^2\, dx + \int_{X_N}^{R} |\psi'(x,t)|^2\, dx$$

$$+ \int_{-R}^{R} |\dot\psi(x,t)|^2\, dx \to 0 \text{ as } t \to +\infty. \tag{1.3.47}$$

It remains to check this convergence with appropriate $a_k(t)$.

1.3.8 Relaxation

To prove (1.3.47), we introduce an appropriate notion of *relaxation*. We define the Sobolev norm of the space $H^1(-R, R)$ as usual:

$$\|z\|_R^2 \equiv \|z'\|_R^2 + \|z\|_R^2. \tag{1.3.48}$$

Definition 1.3.19 (i) A function $z(t) \in L^2_{\text{loc}}(\mathbb{R}^+)$ is called relaxing in L^2 if there exists a function $\bar z(t)$ such that for every $R > 0$,

$$\|z(\cdot + t) - \bar z(t)\|_R^2 \to 0 \quad \text{as} \quad t \to +\infty. \tag{1.3.49}$$

We denote this by $z(t) \overset{L^2}{\sim} \bar z(t)$ as $t \to +\infty$.
(ii) A function $z(t) \in H^1_{\text{loc}}(\mathbb{R}^+)$ is called relaxing in H^1 if there exists a function $\bar z(t)$ such that for every $R > 0$,

$$\|z(\cdot + t) - \bar z(t)\|_R \to 0 \quad \text{as} \quad t \to +\infty. \tag{1.3.50}$$

We denote this relation by $z(t) \overset{H^1}{\sim} \bar z(t)$ as $t \to +\infty$.

The following properties of the relaxation are evident.

R0. We may assume $\bar z(t) \equiv z(t)$ in (1.3.50) without loss of generality.

R1. If the function $z(t)$ is relaxing in H^1, then it is relaxing in L^2 as well.

R2. For the function $z(t)$ to be relaxing in L^2, it suffices that

$$\int_0^\infty |z(t)|^2 dt < \infty. \tag{1.3.51}$$

In this case we may set $\bar z(t) \equiv 0$, i.e., $z(t) \overset{L^2}{\sim} 0$ as $t \to +\infty$.

R3. For the function $z(t)$ to be relaxing in H^1, it suffices that

$$\int_0^\infty |z'(t)|^2 dt < \infty. \tag{1.3.52}$$

Indeed, (1.3.52) implies by the Cauchy–Schwarz inequality for $|x| \le R$ that

$$|z(x+t) - z(t)| = \left| \int_t^{x+t} z'(s)\, ds \right| \le R^{1/2} \|z'(\cdot + t)\|_R^{1/2} \to 0 \text{ as } t \to +\infty. \tag{1.3.53}$$

R4. If the function $z(t)$ is relaxing in H^1, then its derivative $z'(t)$ is relaxing in L^2, and $z'(t) \overset{L^2}{\sim} 0$ as $t \to +\infty$ according to **R2**.

R5. Conversely, if $z(t)$ is relaxing in L^2, then the integral $y(t) \equiv \int_{t+h_-}^{t+h_+} z(s)\, ds$ is relaxing in H^1 for any $h_\pm \in \mathbb{R}$, and we may take

$$\overline{y}(t) \equiv (h_+ - h_-)\overline{z}(t). \tag{1.3.54}$$

R6. If $z(t) \sim \overline{z}(t)$ as $t \to +\infty$ in L^2 (or in H^1), then $z(t+h) \sim \overline{z}(t)$ in L^2 (or in H^1) for every $h \in \mathbb{R}$.

R7. The set of all functions $z(t)$ relaxing in L^2 (or in H^1) is a vector space, and $z_1(t) + z_2(t) \sim \overline{z}_1(t) + \overline{z}_2(t)$, if $z_j(t) \sim \overline{z}_j(t)$, $j = 1, 2$.

R8. Let $F(\cdot) \in C^1(\mathbb{R})$ and $y(t) \in C_b(\mathbb{R}^+)$. Then $y(t) \overset{L^2}{\sim} \overline{y}(t)$ implies $F(y(t)) \overset{L^2}{\sim} F(\overline{y}(t))$.

In the next section we establish the relaxation of the Cauchy data of the solution $\psi(x,t)$ on the lines $x = x_k \pm 0$,

$$y_k(t) \equiv \psi(x_k, t) \text{ and } z_k^{\pm}(t) \equiv \psi'(x_k \pm 0, t), \quad t \in \mathbb{R}, \quad k = 1, \dots, N. \tag{1.3.55}$$

Lemma 1.3.20 *All the functions $y_k(t)$, $k = 1, \dots, N$, are relaxing in H^1 and all the functions $z_k^{\pm}(t)$, $k = 1, \dots, N$, are relaxing in L^2. Moreover,*
$$\dot{y}_1, \dot{y}_N, z_1^-, z_N^+ \overset{L^2}{\sim} 0 \text{ as } t \to +\infty.$$

Let us show that this lemma, the d'Alembert representation (1.3.9), and the properties **R0–R8** of the relaxation imply (1.3.47). The d'Alembert representation (1.3.9) leads to the well-known d'Alembert formula for $x_k < x < x_{k+1}$ with $k \ge 1$:

$$\psi(x,t) = \frac{y_k(t - (x - x_k)) + y_k(t + (x - x_k))}{2} + \frac{1}{2} \int\limits_{t-(x-x_k)}^{t+(x-x_k)} z_k^+(s)\,ds.$$

$$(1.3.56)$$

Therefore,

$$\psi'(x,t) = \frac{-y_k'(t - (x - x_k)) + y_k'(t + (x - x_k))}{2}$$

$$+ \frac{z_k^+(t + (x - x_k)) + z_k^+(t - (x - x_k))}{2}. \qquad (1.3.57)$$

Hence, Lemma 1.3.20 and **R4–R7** imply (1.3.47) with $a_k(t) = \overline{z_{k-1}^+}(t)$ for $k = 2, \ldots, N$.

1.3.9 Scattering of Energy to Infinity

Here we analyze the energy scattering to infinity, which will be applied for the proof of Lemma 1.3.20 in the next section.

Lemma 1.3.21 *The following bound holds:*

$$\int\limits_0^\infty (|\dot{y}_1(t)|^2 + |z_1^-(t)|^2 + |\dot{y}_N(t)|^2 + |z_N^+(t)|^2)\,dt < \infty. \qquad (1.3.58)$$

Proof The d'Alembert representations (1.3.9) with $k = 1$ and $k = N+1$ imply that (1.3.58) is equivalent to

$$\int\limits_0^\infty (|f_1'(x_1 - t)|^2 + |g_1'(x_1 + t)|^2 + |f_{N+1}'(x_N - t)|^2$$

$$+ |g_{N+1}'(x_N + t)|^2)\,dt < \infty. \qquad (1.3.59)$$

The integrals for the *incident waves* $f_1'(x_1 - t)$ and $g_{N+1}'(x_N + t)'$ are finite due to the d'Alembert formulas (1.2.24)

$$f_1'(x) = \frac{\psi_0'(x)}{2} - \frac{1}{2}\pi_0(x), \quad x < x_1,$$

$$g_{N+1}'(x) = \frac{\psi_0'(x)}{2} + \frac{1}{2}\pi_0(x), \quad x > x_N,$$

where $(\psi_0, \pi_0) := Y(0) \in \mathcal{E}$. To derive (1.3.59) for g_1', f_{N+1}', we introduce the energy functional for $Y = (\psi(x), \pi(x)) \in \mathcal{E}$ in the interval $\Delta = [x_1, x_N]$:

$$\mathcal{H}_\Delta(Y) = \frac{1}{2} \int_{x_1}^{x_N} \left[|\pi(x)|^2 + |\psi'(x)|^2 \right] dx + \sum_{k=1}^{N} V_k(y_k), \quad \text{where} \ \ y_k = \psi(x_k).$$

(1.3.60)

Let us calculate the energy flow from Δ: (1.3.1) and (1.3.9) with $k = 1$ and $N + 1$ imply for initial data $(\psi_0, \pi_0) \in \mathcal{E}$ that

$$\frac{d}{dt} \mathcal{H}_\Delta(Y(t)) = \dot\psi \psi' \Big|_{x=x_1-0}^{x=x_N+0}$$

$$= |f_1'(x_1 - t)|^2 - |g_1'(x_1 + t)|^2 + |g_{N+1}'(x_N + t)|^2$$

$$- |f_{N+1}'(x_N - t)|^2. \tag{1.3.61}$$

Integrating, we get the energy balance,

$$\mathcal{H}_\Delta(Y(t)) + \int_0^t (|g_1'(x_1 + s)|^2 + |f_{N+1}'(x_N - s)|^2) \, ds$$

$$= \mathcal{H}_\Delta(Y(0)) + \int_0^t (|f_1'(x_1 - s)|^2 + |g_{N+1}'(x_N + s)|^2) \, ds, \quad t \in \mathbb{R}.$$

(1.3.62)

Then the bounds (1.3.59) for g_1', f_{N+1}' follow from the same bounds (1.3.59) for f_1', g_{N+1}' because $\inf_{Y \in \mathcal{E}} \mathcal{H}_\Delta(Y) > -\infty$ due to the middle condition of (1.3.6). $\qquad\square$

Remark 1.3.22 The integral of the RHS of (1.3.61) over time interval $[0, t]$ is the energy radiated outside (cf. Remark 1.2.8).

1.3.10 Proof of Relaxation

We prove Lemma 1.3.20 by induction in k.

ad $k = 1$ *and* $k = N$. (1.3.58) implies the needed relaxation of $y_1(t)$, $y_N(t)$ and of $z_1^-(t)$, $z_N^+(t)$ according to **R3** and to **R2**, respectively. Then the relaxation of $z_1^+(t)$ and $z_{N+1}^-(t)$ follows by **R7** and **R8** from the third equation of (1.3.1) with $k = 1, N$, that is,

$$z_k^+(t) - z_k^-(t) = -F_k(y_k(t)), \quad t \in \mathbb{R}, \tag{1.3.63}$$

taking into account the estimates (1.3.16).

ad k = 2 Let us prove the relaxation of $y_2(t)$ and $z_2^-(t)$. First, (1.3.56) with $k = 2$ and $x = x_2$ implies

$$y_2(t) = \psi(x_2, t) = \frac{y_1(t - l_2) + y_1(t + l_2)}{2}$$

$$+ \frac{1}{2} \int_{t-l_2}^{t+l_2} z_1^+(s) \, ds, \quad l_2 := |x_2 - x_1|. \qquad (1.3.64)$$

Therefore, **R5** and **R6** imply the relaxation of $y_2(t)$ in H^1. At last we take derivatives in (1.3.56) and get

$$z_2^-(t) \equiv \psi'(x_2 - 0, t) = \frac{-\dot{y}_1(t - l_2) + \dot{y}_1(t + l_2)}{2} + \frac{z_1^+(t + l_2) + z_1^+(t - l_2)}{2}.$$
$$(1.3.65)$$

Therefore, **R2**, **R6**, and **R7** imply the relaxation of $z_2^-(t)$ in L^2. The proof of Lemma 1.3.20 can be completed by induction.

1.4 Space-Localized Nonlinearity

In this section we present the result [43] on global attraction to stationary states for nonlinear wave equations with general nonlinearity

$$\ddot{\psi}(x, t) = \psi''(x, t) + f(x, \psi(x, t)), \qquad x \in \mathbb{R}, \qquad (1.4.1)$$

where $f(x, \psi) = \chi(x) F(\psi)$, $F(\psi) = -\nabla U(\psi)$ for $\psi \in \mathbb{R}^d$, and

$$U(\psi) \in C^2(\mathbb{R}^d), \quad \chi \in C_0^\infty(\mathbb{R}), \qquad (1.4.2)$$

$$\chi(x) \geq 0, \quad \chi(x) \not\equiv 0. \qquad (1.4.3)$$

We will consider the Cauchy problem for equation (1.4.1) with initial conditions

$$\psi(x, 0) = \psi_0(x), \quad \dot{\psi}(x, 0) = \pi_0(x), \qquad x \in \mathbb{R}. \qquad (1.4.4)$$

The equation (1.4.1) can be written as the dynamical system

$$\dot{Y}(t) = \mathbf{F}(Y(t)), \qquad t \in \mathbb{R}, \qquad (1.4.5)$$

with $Y(t) = (\psi(t), \dot{\psi}(t))$. This equation also can be written as the Hamiltonian system (1.1.2) with Hamiltonian functional

$$\mathcal{H}(\psi, \pi) = \frac{1}{2} \int [|\pi(x)|^2 + |\psi'(x)|^2 + \chi(x)U(\psi(x, t))] \, dx, \qquad (\psi, \pi) \in \mathcal{E}, \qquad (1.4.6)$$

where the Hilbert phase space $\mathcal{E}$ is defined in Definition 1.3.1. We assume that the potential U is confining, i.e.,

$$U(\psi) \to \infty, \qquad |\psi| \to \infty. \tag{1.4.7}$$

Denote by $\mathcal{E}_F$ the space $\mathcal{E}$ endowed with seminorms (1.3.5).

Proposition 1.4.1 *Let $d \geq 1$ and assumptions (1.4.2), (1.4.3), and (1.4.7) hold. Then*

(i) For every initial state $Y(0) \in \mathcal{E}$, equation (1.4.5) has a unique solution $Y(t) \in C(\mathbb{R}, \mathcal{E})$.

(ii) The mapping $W(t) : Y(0) \mapsto Y(t)$ is continuous in $\mathcal{E}$ and in $\mathcal{E}_F$ for every $t \in \mathbb{R}$.

(iii) The energy (1.4.6) is conserved,

$$\mathcal{H}(Y(t)) = \text{const}, \qquad t \in \mathbb{R}. \tag{1.4.8}$$

Definition 1.4.2 $\mathcal{S}$ *denotes the set of all stationary states $S = (s(x), 0) \in \mathcal{E}$ for the equation (1.4.1).*

The functions $s(x)$ satisfy the stationary equation

$$s''(x) + f(x, s(x)) = 0, \qquad x \in \mathbb{R}. \tag{1.4.9}$$

The next proposition gives a criterion for the set $\mathcal{S}$ being a nonempty discrete subset of the space $\mathcal{E}_F$. Denote by $\mathcal{U}$ the potential energy functional:

$$\mathcal{U}(\psi) := \mathcal{H}(\psi, 0) = \int_{\mathbb{R}} \left[\frac{1}{2}|\psi'(x)|^2 + \chi(x)U(\psi(x)) \right] dx, \qquad \psi \in E_c. \tag{1.4.10}$$

Proposition 1.4.3 *Let conditions (1.4.2), (1.4.3), and (1.4.7) hold, and moreover, let $d = 1$ and the function $F(\psi)$ be real-analytic on $\mathbb{R}$. Then $\mathcal{S}$ is a discrete subset of $\mathcal{E}_F$.*

The main result of [43] is the following theorem, which is illustrated by Figure 1.

Theorem 1.4.4 *(i) Let conditions (1.4.2), (1.4.3), and (1.4.7) hold and $Y(0) \in \mathcal{E}$. Then the corresponding solution $Y(t) = (\psi(t), \pi(t)) \in C(\mathbb{R}, \mathcal{E})$ to equation (1.4.5) attracts to $\mathcal{S}$ in the sense (1.2.18):*

$$Y(t) \xrightarrow{\mathcal{E}_F} \mathcal{S}, \qquad t \to \pm\infty. \tag{1.4.11}$$

(ii) Suppose additionally that $d = 1$ and that the function $F(\psi)$ is real-analytic for $\psi \in \mathbb{R}$. Then, for any solution $Y(t) = (\psi(t), \pi(t)) \in C(\mathbb{R}, \mathcal{E})$ to equation (1.4.5),

$$Y(t) \xrightarrow{\mathcal{E}_F} S_\pm \in \mathcal{S}, \qquad t \to \pm\infty. \tag{1.4.12}$$

Remarks 1.4.5 (i) The assertion (ii) of this theorem follows from (i) due to Proposition 1.4.3.

(ii) The convergences (1.4.11) and (1.4.6), (1.4.2), (1.4.3), (1.4.7) imply (10) by the Fatou theorem.

1.4.1 Plan of the Proof

It suffices to consider only the case $t \to \infty$. Our proofs of global attraction (1.4.11) and (1.4.12) rely on a novel method of *omega-limit trajectories*, which is a development of the method of omega-limit points used in [42]; see Section 1.3. Later on, this method played a central role in the theory of global attractors for $U(1)$-invariant PDEs [60]–[71].

By (1.4.2), we have

$$\operatorname{supp} \chi \subset \Delta := [-a, a] \tag{1.4.13}$$

for some $a > 0$. Conditions (1.4.3) and (1.4.7) imply the finiteness of the energy radiated from the segment Δ. Hence, similarly to (1.3.58),

$$\int_0^\infty [|\dot\psi(-a,t)|^2 + |\psi'(-a,t)|^2 + |\dot\psi(a,t)|^2 + |\psi'(a,t)|^2]dt < \infty. \tag{1.4.14}$$

This means, roughly, that

$$\psi(\pm a, t) \sim C_\pm, \qquad \psi'(\pm a, t) \sim 0, \qquad t \to \infty. \tag{1.4.15}$$

More precisely, the functions $\psi(\pm a, t)$ and $\psi'(\pm a, t)$ are slowly varying for large times, so their shifts form compact families. Namely, from an arbitrary sequence $s_k \to \infty$, one can choose a subsequence $s_{k'} \to \infty$ such that *for any* $T > 0$, the following *uniform* convergence holds:

$$\psi(\pm a, t + s_{k'}) \to C_\pm \quad \text{for} \quad t \in [0,T], \qquad k' \to \infty, \tag{1.4.16}$$

where the constants $C_\pm$ depend on the subsequence. It remains to prove that *for any* $T > 0$,

$$\psi(x, t + s_{k'}) \to S_+(x) \in \mathcal{S} \quad \text{for} \quad t \in [0,T] \text{ and } x \in [-a,a], \qquad k' \to \infty, \tag{1.4.17}$$

where the convergence holds in $C([0,T]; H^1[-a,a])$. In other words, *each omega-limit trajectory is a stationary state.*

To deduce (1.4.17) from (1.4.16), we need, roughly speaking, to justify the well-posedness of the boundary value problem for a nonlinear differential equation (1.4.1) in the half-strip $-a \le x \le a$, $t > 0$, with the Cauchy boundary conditions (1.4.15) on the sides $x = \pm a$. Then the convergence (1.4.16) of boundary values implies the convergence (1.4.17) of the solution inside the strip.

Our main idea is to use evident symmetry of the wave equation with respect to interchange of variables x and t with a simultaneous change of the sign of the potential U; that is, (1.4.1) can be written as

$$\psi''(x,t) = \ddot{\psi}(x,t) - f(x,\psi(x,t)). \qquad (1.4.18)$$

However, in this equation, with the "time" x, the condition (1.4.7) makes new potential $-U$ unbounded from below! Consequently, these dynamics with x as the time variable are not correct on the interval $|x| \le a$.

For example, in the case $U(\psi) = \psi^4$, the equation (1.4.1) for solutions of type $\psi(x,t) = \psi(x)$ is $\psi''(x) - 4\psi^3(x) = 0$. Solutions of this ordinary differential equation with finite Cauchy initial data at $x = -a$ can become infinite at any point $x \in (-a,a)$. However, in our situation, local well-posedness is sufficient due to *a priori bounds*, which follow from the energy conservation (1.4.8) in view of the conditions (1.4.2), (1.4.3), and (1.4.7).

Remark 1.4.6 The discreteness of the set S is essential for the asymptotics (1.4.12). For example, convergence (1.4.12) fails for the solution $\psi(x,t) = \sin[\log(|x-t|+2)]$ in the case when $d=1$ and $F(\psi) = 0$ for $|\psi| \le 1$.

1.4.2 Well-Posedness and A Priori Estimates

Proposition 1.4.1 follows by classical technique [12]. The energy conservation (1.3.13) implies a priori estimates

$$\sup_{t\in\mathbb{R}} \|Y(t)\|_{\mathcal{E}} < \infty \qquad (1.4.19)$$

due to the conditions (1.4.3) and (1.4.7). We need, however, a finer characterization of the properties of the solutions.

Proposition 1.4.7 *Let the assumptions (1.4.2), (1.4.3), and (1.4.7) hold. Then*

(i) The mapping $W(t)$ is Lipschitz-continuous in $\mathcal{E}_F$, and for every $R,T > 0$,

$$\|W(t)Y_1 - W(t)Y_2\|_R \le L_T \|Y_1 - Y_2\|_{R+T} \text{ for } |t| \le T, \qquad (1.4.20)$$

where L_T is bounded for bounded norms $\|Y_1\|_{R+T}, \|Y_2\|_{R+T}$.

(ii) For solutions $Y(t) = (\psi(t),\dot{\psi}(t)) \in C(\mathbb{R},\mathcal{E})$, the a priori estimate holds:

$$|\psi(x,t)| \le b(x) := \alpha + \beta\sqrt{|x|}, \qquad (x,t) \in \mathbb{R}^2, \qquad (1.4.21)$$

where α and β are bounded for bounded energy $\mathcal{H}(Y(0))$.

(iii) $\psi(x, \cdot)$ *is a continuous function of* $x \in \mathbb{R}$ *with values in* $H^1_{\text{loc}}(\mathbb{R})$, *and* $\psi'(x, \cdot)$ *is a continuous function of* $x \in \mathbb{R}$ *with values in* $L^2_{\text{loc}}(\mathbb{R})$.

(iv) For a.a. $x \in \mathbb{R}$ *and any* $t \in \mathbb{R}$,

$$\int_t^{t+1} (|\dot{\psi}(x,\tau)|^2 + |\psi'(x,\tau)|^2 + |\psi(x,\tau)|^2)d\tau \le e < \infty. \qquad (1.4.22)$$

Proof ad (i) For solutions $Y(t) \in C(\mathbb{R}, \mathcal{E})$ to the nonlinear equation (1.4.1), the Duhamel representation holds (see [3]):

$$W(t)Y(0) = W_0(t)Y(0)$$

$$+ \int_0^t W_0(t-s)f_*(\cdot,s)ds, \quad f_*(x,s) := (0, f(x, \psi(x,s))).$$

$$(1.4.23)$$

Here $W_0(t)$ denotes the dynamical group corresponding to the linear equation (1.4.1) with $f(x,u) \equiv 0$, i.e.,

$$W_0(t)(\psi(0), \pi(0)) = (\psi(t), \dot{\psi}(t)),$$

where $\psi(x,t)$ is given by the d'Alembert formula (1.1.6). This formula implies the Lipschitz continuity (1.4.20) for $W_0(t)$. Then, for $W(t)$, the same continuity follows from (1.4.23) by (1.4.2) and (1.4.19).

ad (ii) The bound (1.4.19) implies that

$$D := \sup_{t \in \mathbb{R}} \int |\psi'(x,t)|^2 dx < \infty, \qquad (1.4.24)$$

and D is bounded for bounded energy $\mathcal{H}(Y(0))$. Therefore, by the Cauchy–Schwarz inequality,

$$|\psi(x,t) - \psi(0,t)| = \left| \int_0^x \psi'(y,t)dy \right| \le \sqrt{D}\sqrt{|x|}, \qquad (x,t) \in \mathbb{R}^2. \quad (1.4.25)$$

At last, $\sup_{t \in \mathbb{R}} |u(0,t)| < \infty$ by the bound (1.4.19). Now (1.4.25) implies (1.4.21).

ad (iii) and (iv) For $|x| > a$ the claimed properties follow similarly to (1.4.14). To prove them for $|x| < a$, rewrite the equation (1.4.1) as (1.4.18) and apply the integral representation of the type (1.4.23),

$$Z(x) = W_0(x+a)Z(-a)$$

$$- \int_{-a}^x W_0(x-y)f_*(y, \cdot)dy, \qquad Z(x) := (\psi(x, \cdot), \psi'(x, \cdot)). \quad (1.4.26)$$

The claimed properties for the first term on the RHS follow from (1.4.14), and for the integral term, these properties follow from (1.4.2) and estimates (1.4.21). □

1.4.3 Stationary States

We prove Proposition 1.4.3 by a suitable modification of the arguments from the proof of Proposition 1.3.5. The stationary equation (1.4.9) and conditions (1.4.2) imply that all stationary solutions $s(x)$ are smooth and

$$s(x) = s(\pm a), \qquad \pm x \geq a, \qquad (1.4.27)$$

since $s'(x) \in L^2(\mathbb{R})$. Hence, the variation $D\mathcal{U}(s)$ exists and

$$D\mathcal{U}(s) = -s''(x) + f(x, s(x)).$$

Therefore, equation (1.4.1) for stationary states implies the variational equation

$$D\mathcal{U}(s) = 0. \qquad (1.4.28)$$

The identities (1.4.27) imply that the continuous map $I : \mathcal{E}_F \to \mathbb{R}$ defined by

$$I(\psi(x), \pi(x)) := \psi(-a)$$

is a homeomorphism on $\mathcal{S}$. Hence, Proposition 1.4.3 obviously follows from the next lemma.

Lemma 1.4.8 $Z := I\mathcal{S}$ *is a discrete subset of* $\mathbb{R}$.

Proof We should prove that Z has no limit points. Let us assume the contrary – that there exists an infinite subsequence

$$z_k \in Z, \qquad z_k \to \bar{z} \in Z, \qquad k \to \infty. \qquad (1.4.29)$$

All stationary states satisfy the boundary value problem

$$\left\{ \begin{array}{ll} s_\lambda''(x) + f(x, s_\lambda(x)) = 0 & x \in [-a, a] \\ s_\lambda(-a) = \lambda, \qquad s_\lambda'(-a) = 0 \end{array} \right. \bigg|. \qquad (1.4.30)$$

Let us denote by Λ the set of all $\lambda \in \mathbb{R}$ such that the solution to (1.4.30) exists. We extend $s_\lambda(x)$ to $|x| > a$ by constants,

$$s_\lambda = s_\lambda(\pm a) \text{ for } \pm x > a. \qquad (1.4.31)$$

Then $S_\lambda = (s_\lambda, 0) \in \mathcal{E}$ for every $\lambda \in \Lambda$, though generally $S_\lambda \notin \mathcal{S}$.

Let us define the map $T : \Lambda \to \mathbb{R}$ by

$$T(\lambda) := s'_\lambda(a - 0). \tag{1.4.32}$$

Then

$$Z = \{\lambda \in \Lambda : \ T(\lambda) = 0\}, \tag{1.4.33}$$

and (1.4.29) implies that

$$T(\bar{z}) = T(z_k) = 0, \qquad k = 1, \dots. \tag{1.4.34}$$

The set Λ is an open subset of $\mathbb{R}$, hence

$$\Lambda = \bigcup_1^\infty \Lambda_j, \tag{1.4.35}$$

where Λ_j are open intervals. We have $\bar{z} \in \Lambda_* = \Lambda_l$ with some l. Let us show that

$$\Lambda_* = \mathbb{R}. \tag{1.4.36}$$

Namely, the map T is real-analytic on Λ_*, and hence, (1.4.29) and (1.4.34) imply that

$$T(\lambda) \equiv 0, \qquad \lambda \in \Lambda_*, \tag{1.4.37}$$

since Λ_* is an open and connected subset of $\mathbb{R}$. Hence, definition (1.4.33) implies that

$$\Lambda_* \subset Z. \tag{1.4.38}$$

Now (1.4.28) implies that

$$\partial_\lambda \mathcal{U}(s_\lambda) = \langle D\mathcal{U}(s_\lambda), \partial_\lambda s_\lambda \rangle = 0, \qquad \lambda \in \Lambda_*. \tag{1.4.39}$$

Hence,

$$\mathcal{U}(s_\lambda) \equiv \mathcal{U}(s_{\bar{z}}), \qquad \lambda \in \Lambda_*. \tag{1.4.40}$$

However, this identity implies that the set $\mathcal{S}_* := \{S_\lambda : \lambda \in \Lambda_*\}$ is bounded in $\mathcal{E}$ by conditions (1.4.2) and (1.4.3). Hence, $\mathcal{S}_*$ is precompact in $C(\mathbb{R}, \mathbb{R} \times \mathbb{R})$. Its closure in $C(\mathbb{R}, \mathbb{R} \times \mathbb{R})$ obviously belongs to $\mathcal{S}$, and hence,

$$\overline{\Lambda_*} \subset \Lambda_*. \tag{1.4.41}$$

Now (1.4.36) follows. Moreover, now (1.4.38) implies that $Z = \mathbb{R}$, which contradicts the boundedness of $\mathcal{S}_*$ in $\mathcal{E}$. This contradiction completes the proof of Lemma 1.4.8. $\qquad\square$

1.4.4 Long-Time Asymptotics

We prove the Theorem 1.4.4.

Compact Attracting Set

Let us construct a compact attracting set $\mathcal{A}$ for the considered trajectory $Y(t)$. Let $b > 0$ denote some constant to be chosen later.

Definition 1.4.9 $\mathcal{A}_b := \{S_\lambda = (s_\lambda(x), 0) \in \mathcal{E} : \lambda \in \Lambda, |s_\lambda(x)| \le b \text{ for } |x| \le a\}.$

$\mathcal{A}_b$ is a compact set in $\mathcal{E}_F$ due to the equation (1.4.9). We prove the next lemma in the following section.

Lemma 1.4.10 *Let assumptions of Theorem 1.4.4 hold. Then*

$$Y(t) \xrightarrow{\mathcal{E}_F} \mathcal{A} = \mathcal{A}_b, \qquad t \to \pm\infty, \qquad (1.4.42)$$

if the constant b is sufficiently large.

Proof of Theorem 1.3 (i)

The next lemma implies the attraction to stationary states (1.4.11).

Lemma 1.4.11 *(i)* $\omega(Y) \ne \emptyset$ *and (ii)* $\omega(Y) \subset \mathcal{S}$.

Proof (i) Lemma 1.4.10 means that there exists a function $\lambda(t) \in C[0, \infty)$ such that for every $R > 0$,

$$\|Y(t) - S_{\lambda(t)}\|_R \to 0 \quad \text{as} \quad t \to +\infty. \qquad (1.4.43)$$

Here $S_{\lambda(t)} = (s_{\lambda(t)}, 0)$, where $s_{\lambda(t)}(x)$ is defined by (1.4.30) and (1.4.31).
    The orbit $\{S_{\lambda(t)} : t > 0\}$ is precompact in $\mathcal{E}_F$. Hence, the limit (1.4.43) implies that the orbit $\{Y(t) : t > 0\}$ is also precompact in $\mathcal{E}_F$. Therefore, $\omega(Y) \ne \emptyset$.
(ii) $\omega(Y) \subset \mathcal{A}$ by (1.4.42). Moreover, the set $\omega(Y)$ is invariant with respect to $W(t)$ due to the continuity of $W(t)$ in $\mathcal{E}_F$. Hence, for every $\overline{Y} \in \Omega(Y)$, there exists a C^1-curve $t \mapsto \lambda(t) \in \mathbb{R}$ such that $W(t)\overline{Y} = S_{\lambda(t)}$. Then $S_{\lambda(t)}$ is the solution to (1.4.5). In particular, $\partial_t S_{\lambda(t)} \equiv 0$. Therefore, $\lambda(t) \equiv \lambda$ and $\overline{Y} = S_\lambda \in \mathcal{S}$.   □

1.4.5 Attraction to a Compact Set

We deduce Lemma 1.4.10 from the following lemma on "attraction in the mean," which we prove in the next section. Let us denote for $R > 0$,

$$\rho_R(t) = \inf_{S \in \mathcal{A}} \|Y(t) - S\|_R \quad \text{for } t \in \mathbb{R}. \qquad (1.4.44)$$

Lemma 1.4.12 *For every $R > 0$,*

$$\int_0^\infty \rho_R^2(t)dt < \infty. \tag{1.4.45}$$

Proof of Lemma 1.4.10 Let us fix a metric $d(\cdot,\cdot)$ on $\mathcal{E}$, defining the topology of $\mathcal{E}_F$. We prove (1.4.42) ad absurdum: let us assume that there exist $\varepsilon > 0$ and a sequence $t_k \to \infty$, such that

$$d(Y(t_k), \mathcal{A}) \geq \varepsilon \text{ for all } k = 1, 2, \ldots. \tag{1.4.46}$$

We will show that this is impossible, and this completes the proof of Lemma 1.4.10. We may assume that $t_k + 1 < t_{k+1}$ for every k. Now (1.4.45) implies by the Fatou theorem that

$$\int_0^1 \sigma_R(\theta)d\theta < \infty, \text{ where } \sigma_R(\theta) = \sum_1^\infty \rho_R^2(t_k + \theta). \tag{1.4.47}$$

Therefore, $\sigma_R(\theta) < \infty$ for every θ in a subset $\Theta(R) \subset [0,1]$ with $\int_{\Theta(R)} dx = 1$. Then, for every $R > 0$,

$$\rho_R(t_k + \theta) \to 0 \text{ as } k \to \infty \text{ for } \theta \in \Theta := \cap_{R\in\mathbb{N}}^\infty \Theta(R). \tag{1.4.48}$$

Hence $Y(t_k + \theta) \xrightarrow{\mathcal{E}_F} \mathcal{A}$ as $k \to \infty$ for every $\theta \in \Theta \subset [0,1]$, and $\int_\Theta dx = 1$. Hence, for every $\theta \in \Theta$, the compactness of $\mathcal{A}$ in $\mathcal{E}_F$ implies that for some sequence $k(\theta) \to \infty$,

$$Y(t_{k(\theta)} + \theta) \xrightarrow{\mathcal{E}_F} \overline{Y}(\theta) \in \mathcal{A} \text{ as } k(\theta) \to \infty, \quad \theta \in \Theta. \tag{1.4.49}$$

Then the continuity of maps $W(-\theta)$ in $\mathcal{E}_F$ implies also that

$$Y(t_{k(\theta)}) \xrightarrow{\mathcal{E}_F} W(-\theta)\overline{Y}(\theta) \text{ as } k(\theta) \to \infty, \quad \theta \in \Theta. \tag{1.4.50}$$

On the other hand, the compactness of $\mathcal{A}$ in $\mathcal{E}_F$ implies that there exists a sequence $\theta_j \in \Theta$ such that $\theta_j \to 0$ as $j \to \infty$ and

$$\overline{Y}(\theta_j) \xrightarrow{\mathcal{E}_F} Y^* \in \mathcal{A} \text{ as } j \to \infty. \tag{1.4.51}$$

Now the uniform Lipschitz continuity (1.4.20) of $W(-\theta)$ with $\theta \in [0,1]$ and the convergence $W(-\theta_j)Y^* \xrightarrow{\mathcal{E}_F} Y^*$ as $j \to \infty$ imply that

$$W(-\theta_j)\overline{Y}(\theta_j) \xrightarrow{\mathcal{E}_F} Y^* \text{ as } j \to \infty. \tag{1.4.52}$$

However, this convergence together with (1.4.50) for $\theta = \theta_j$ contradicts (1.4.46). $\square$

1.4.6 Attraction in the Mean

We prove Lemma 1.4.12. It suffices to construct for sufficiently large $b > 0$ a function $S_{\mu(t)} = (s_{\mu(t)}, 0) \in \mathcal{A}_b$ defined for $t \geq T$ with sufficiently large $T > 0$ such that for every $R > 0$,

$$\int_T^\infty \|Y(t) - S_{\mu(t)}\|_R^2 dt < \infty. \tag{1.4.53}$$

We will establish this inequality with

$$\mu(t) = y_-(n) := \psi(-a, n), \qquad n \leq t < n + 1, \tag{1.4.54}$$

where $n = 0, 1, \ldots$ and $n \geq T$. We may change the seminorm $\|\cdot\|_R$ from (1.3.5) by an equivalent seminorm with $|\psi(-a)|$ instead of $|\psi(0)|$. Then (1.4.53) means for $R > a$ that

$$\int_T^\infty \left(\int_{|x|<a} (|\psi'(x,t) - s'_{\mu(t)}(x)|^2 + |\psi(x,t) - s_{\mu(t)}(x)|^2 + |\dot\psi(x,t)|^2) dx \right.$$

$$\left. + |\psi(-a,t) - \mu(t)|^2 + \int_{a<|x|<R} (|\psi'(x,t)|^2 + |\dot\psi(x,t)|^2) dx \right) dt < \infty. \tag{1.4.55}$$

Energy Scattering to Infinity

The bound (1.4.14) can be written as

$$\int_0^\infty (|\dot y_-(t)|^2 + |z_-(t)|^2 + |\dot y_+(t)|^2 + |z_+(t)|^2) \, dt < \infty, \tag{1.4.56}$$

where $y_\pm(t) = \psi(\pm a, t)$ and $z_\pm(t) = \psi'(\pm a, t)$. This follows similarly to (1.3.58) from d'Alembert representation

$$\psi(x,t) = f_\pm(t - x) + g_\pm(t + x), \qquad \pm x > a, \quad t \in \mathbb{R}, \tag{1.4.57}$$

and from finiteness of the energy flow from the segment $\Delta := [-a, a]$, differentiating the energy functional

$$\mathcal{H}_\Delta(Y) = \int_\Delta \left[\frac{|\pi(x)|^2}{2} + \frac{|\psi'(x)|^2}{2} + V(x, \psi(x)) \right] dx,$$

$$Y = (\psi(x), \pi(x)) \in \mathcal{E}. \tag{1.4.58}$$

Now (1.4.56) implies that

$$\int\limits_0^\infty |\psi(-a,t) - \mu(t)|^2 dt = \int\limits_0^\infty |y_-(t) - y_-([t])|^2 dt$$

$$\leq \sum_{n=0}^\infty \int\limits_0^1 |\dot{y}_-(n+s)|^2 ds < \infty. \qquad (1.4.59)$$

Furthermore, similarly to (1.4.56),

$$\int\limits_0^\infty (|\dot{\psi}(x,t)|^2 + |\psi'(x,t)|^2\,dt \leq C < \infty, \qquad a < |x| < R. \qquad (1.4.60)$$

Hence, the last integral of (1.4.55) is finite. It remains to prove the finiteness of the first integral of (1.4.55),

$$\int\limits_T^\infty \left(\int\limits_{|x|<a} (|\psi'(x,t) - s'_{\mu(t)}(x)|^2 + |\psi(x,t) - s_{\mu(t)}(x)|^2 + |\dot{\psi}(x,t)|^2) dx \right) dt < \infty$$

$$(1.4.61)$$

for sufficiently large $T > 0$. We will deduce (1.4.61) from (1.4.56) in the next section.

Nonlinear Goursat Problem

We consider the Goursat problem for the nonlinear wave equation (1.4.18) with the Cauchy data on the lines $x = $ const:

$$\begin{cases} \phi''(x,t) = \ddot{\phi}(x,t) + f(x,\phi(x,t)), \\ \phi|_{x=r} = u(t), \quad \phi'|_{x=r} = v(t) \end{cases} \Bigg|, \; t \in \mathbb{R}, \quad x \in [r,r+\varepsilon], \quad (1.4.62)$$

where $\varepsilon > 0$. Our assumptions (1.4.2), (1.4.3) provide that the Cauchy problem (1.4.1), (1.4.4) is well posed globally in t. On the other hand, the nonlinear Goursat problem (1.4.62) generally is not well posed globally in $x \in \mathbb{R}$.

We will establish a Lipschitz continuity of the maps

$$G(r,x) : (u(\cdot),v(\cdot)) \mapsto (\phi(x,\cdot),\phi'(x,\cdot)), \qquad x \in [r,r+\varepsilon]$$

in suitable norms for initial data $(u(\cdot),v(\cdot))$ close to $(\psi(r,\cdot),\psi'(r,\cdot))$, where $\varepsilon > 0$ *does not depend on* $r \in [-a,a]$. This continuity holds "along" the considered global solution $\psi(x,t)$ due to the a priori bounds (1.4.19). Using this continuity, we will deduce (1.4.61) from (1.4.56).

Let σ denote an arbitrary segment in $\mathbb{R}$ of the length $|\sigma|$.

Definition 1.4.13 $\mathcal{E}(\sigma) := H^1(\sigma) \oplus L^2(\sigma)$, is the Hilbert space of functions $(u(t), v(t))$ with the norm

$$\|(u,v)\|_{\mathcal{E}(\sigma)} = \|\dot{u}\|_{L^2(\sigma)} + \|u\|_{L^2(\sigma)} + \|v\|_{L^2(\sigma)} < \infty. \tag{1.4.63}$$

Now propositions 1.4.7 (iii) and (iv) imply that for any segment $\sigma \subset \mathbb{R}$,

$$Z_\sigma(r) := (\psi(r,\cdot), \psi'(r,\cdot))|_\sigma \in \mathcal{E}(\sigma) \qquad \text{for a.a. } r \in \mathbb{R}$$

and

$$\|Z_\sigma(r)\|^2_{\mathcal{E}(\sigma)} \le Ce|\sigma| \qquad \text{for a.a. } r \in \mathbb{R}. \tag{1.4.64}$$

Let $\phi_j(x,t)$ with $j = 1,2$ be two solutions of the nonlinear Goursat problem (1.4.62) for $x \in [r, r+\varepsilon)$, where $\varepsilon > 0$, such that $X_j(x) := (\phi(x,\cdot), \phi'(x,\cdot)) \in C(r, r+\varepsilon; H^1_{\text{loc}}(\mathbb{R}) \oplus L^2_{\text{loc}}(\mathbb{R}))$. For such solutions the Goursat problem (1.4.62) is equivalent to the integral identity of type (1.4.26),

$$X_j(x) = W_0(x-r)X_j(r) + \int_r^x W_0(x-y)(0, f(y, \phi_j(y,\cdot))dy, \quad x \in [r, r+\varepsilon]. \tag{1.4.65}$$

For any segment $\sigma = [t_1, t_2]$ and small $\varepsilon > 0$, denote $\sigma_\varepsilon := [t_1 + \varepsilon, t_2 - \varepsilon]$.

Lemma 1.4.14 *Let assumptions (1.4.2), (1.4.3) hold, and*

$$\max_{x \in [r, r+\varepsilon], t \in \mathbb{R}} |\phi_j(x,t)| \le B < \infty, \qquad j = 1, 2. \tag{1.4.66}$$

Then, for any segment $\sigma \subset \mathbb{R}$ with $|\sigma| > 2\varepsilon$,

$$\|X_1(x) - X_2(x)\|_{\mathcal{E}(\sigma_{|x-r|})} \le L(B)\|X_1(r) - X_2(r)\|_{\mathcal{E}(\sigma)}, \qquad x \in [r, r+\varepsilon), \tag{1.4.67}$$

where the Lipschitz constant $L(B)$ does not depend on the segment σ.

Proof By conditions (1.4.2),

$$M(B) := \max_{x \in \mathbb{R}, |\psi| \le B} |\nabla_\psi f(x,\psi)| < \infty. \tag{1.4.68}$$

Hence,

$$\|f(y, \phi_1(y,\cdot)) - f(y, \phi_2(y,\cdot))\|_{L^2(\sigma_{y-r})}$$
$$\le M(B)\|X_1(y) - X_2(y)\|_{\mathcal{E}(\sigma)}, \, y \in [r, r+\varepsilon). \tag{1.4.69}$$

Moreover, the dynamical group $W_0(y)$ admits classical estimate

$$\|W_0(z)X\|_{\mathcal{E}(\sigma_z)} \le \|X\|_{\mathcal{E}(\sigma)}, \qquad z \in [0, |\sigma|/2), \quad X \in \mathcal{E}(\sigma).$$

Now the integral equation (1.4.65) implies the integral inequality

$$m(x) \leq m(r) + M(B) \int_r^x m(y)dy, \qquad x \in [r, r + \varepsilon), \qquad (1.4.70)$$

where

$$m(x) := \|X_1(x) - X_2(x)\|_{\mathcal{E}(\sigma_{|x-r|})}.$$

Hence, the bounds (1.4.67) follow by the Gronwall inequality. $\qquad \square$

Now we can prove the existence of solutions of (1.4.30).

Lemma 1.4.15 *For sufficiently large $t > 0$ the problem (1.4.30) with $\lambda = y_-(t)$ admits a unique solution $s_\lambda(x)$.*

Proof First, the solution exists for $x \in [-a, -a + \varepsilon]$ with sufficiently small $\varepsilon > 0$, which depends on t. This local solution can be extended to all $x \in [-a, a]$ if a priori bounds hold:

$$|s_\lambda(x)| \leq C, \qquad x \in [-a, a]. \qquad (1.4.71)$$

These bounds follow for $\lambda = \lambda(t) := y_-(t)$ with large t by application of Lemma 1.4.14 to the following two solutions $\phi_j(x, t)$ of the nonlinear Goursat problem (1.4.62) with $r = -a$:

$$\phi_1(x, t) := \psi(x, t), \qquad \phi_2(x, t) := s_\lambda(x).$$

We will prove the bounds (1.4.71) with any $C > B_0$, where

$$B_0 := \max_{x \in [-a, a], t \in \mathbb{R}} |\psi(x, t)|. \qquad (1.4.72)$$

The key fact is the following convergence of the Cauchy data of the two solutions

$$\|(\psi(-a, t + \cdot), \psi'(-a, t + \cdot)) - (s_{\lambda(t)}(-a), 0)\|_{\mathcal{E}(\sigma)} \to 0, \qquad t \to \infty \qquad (1.4.73)$$

for any segment $\sigma \subset \mathbb{R}$. This convergence follows from (1.4.56). Now Lemma 1.4.14 implies that

$$\|(\psi(-a + \varepsilon, t + \cdot), \psi'(-a, t + \cdot))$$
$$- (s_{\lambda(t)}(-a + \varepsilon), s'_{\lambda(t)}(-a + \varepsilon))\|_{\mathcal{E}(\sigma_\varepsilon)} \to 0, \qquad t \to \infty$$

if we take $|\sigma| > 2\varepsilon$. Hence, by the Sobolev embedding theorem,

$$\|\psi(-a + \varepsilon, t + \cdot) - s_{\lambda(t)}(-a + \varepsilon)\|_{C(\sigma_\varepsilon)} \to 0, \qquad t \to \infty. \qquad (1.4.74)$$

Therefore, the a priori bounds (1.4.71) hold. $\qquad \square$

Proof of the Attraction in the Mean

Now (1.4.61) follows by the same arguments. Namely, Lemma 1.4.14 implies that for any $\varepsilon \in [0, 2a]$,

$$\|(\psi(-a+\varepsilon, t+\cdot), \psi'(-a+\varepsilon, t+\cdot)) - (s_{\mu(t)}(-a+\varepsilon), s'_{\mu(t)}(-a+\varepsilon))\|_{\mathcal{E}(\sigma_\varepsilon)}$$

$$\leq L(C)\|(\psi(-a, t+\cdot), \psi'(-a, t+\cdot)) - (s_{\mu(t)}(-a), 0)\|_{\mathcal{E}(\sigma)}, \qquad t \geq T$$

for sufficiently large $T > 0$ and $\mu(t)$ defined by (1.4.54). Hence,

$$\|\psi(-a+\varepsilon, t+\cdot) - s_{\mu(t)}(-a+\varepsilon)\|^2_{H^1(\sigma_\varepsilon)}$$

$$+ \|\psi'(-a+\varepsilon, t+\cdot) - s'_{\mu(t)}(-a+\varepsilon)\|^2_{L^2(\sigma_\varepsilon)}$$

$$\leq L(C)[\|\psi(-a, t+\cdot) - s_{\mu(t)}(-a)\|^2_{H^1(\sigma)} + \|\psi'(-a, t+\cdot)\|^2_{L^2(\sigma)}], \quad t \geq T.$$
$$(1.4.75)$$

Choosing here $t = n$, $\sigma_\varepsilon = [0, 1]$ and summing up over $n \geq N \geq T$, we obtain

$$\int_N^\infty (|\dot\psi(x,t)|^2 + |\psi(x,t) - s_{\mu(t)}(x)|^2$$

$$+ |\psi'(x,t) - s'_{\mu(t)}(x)|^2)dt < \infty, \qquad x \in [-a, a] \qquad (1.4.76)$$

since the sum of the right-hand sides is finite by (1.4.56). Moreover, this last sum is bounded, and hence, integrating over $x \in [-a, a]$, we obtain (1.4.61).

Now Lemma 1.4.12 is proved.

1.5 Wave–Particle System

In [44], the first result on global attraction to stationary states (7) is obtained for a 3D real scalar wave field coupled to a relativistic particle. The scalar field satisfies 3D wave equation

$$\ddot\psi(x,t) = \Delta\psi(x,t) - \rho(x - q(t)), \qquad x \in \mathbb{R}^3, \qquad (1.5.1)$$

where $\rho \in C_0^\infty(\mathbb{R}^3)$ is a fixed function representing the charge density of the particle and $q(t) \in \mathbb{R}^3$ is the particle position. The particle motion obeys the Hamiltonian equation with relativistic kinetic energy $\sqrt{1 + p^2}$:

$$\dot q(t) = \frac{p(t)}{\sqrt{1 + p^2(t)}}, \qquad \dot p(t) = -\nabla V(q(t)) - \int \nabla\psi(x,t)\rho(x - q(t))\,dx.$$
$$(1.5.2)$$

Here $-\nabla V(q)$ is an external force corresponding to real potential $V(q)$, and the integral term is a self-force. Thus, wave function ψ is generated by a charged particle and plays the role of a potential acting on the particle, along with the external potential $V(q)$.

The system (1.5.1)–(1.5.2) was introduced by H. Spohn; see [52] for discussion of the physical relevance of this model. This system can formally be represented in Hamiltonian form

$$\dot{\psi} = D_\pi \mathcal{H}, \quad \dot{\pi} = -D_\psi \mathcal{H}, \quad \dot{q}(t) = D_p \mathcal{H}, \quad \dot{p} = -D_q \mathcal{H} \qquad (1.5.3)$$

with Hamiltonian (energy)

$$\mathcal{H}(\psi, \pi, q, p) = \frac{1}{2} \int [|\pi(x)|^2 + |\nabla \psi(x)|^2] \, dx$$

$$+ \int \psi(x)\rho(x - q) \, dx + \sqrt{1 + p^2} + V(q). \qquad (1.5.4)$$

By $\| \cdot \|$, we denote the norm in the Hilbert space $L^2 := L^2(\mathbb{R}^3)$, and $\| \cdot \|_R$ denotes the norm in $L^2(B_R)$, where B_R is the ball $|x| \le R$. Let $\overset{\circ}{H}{}^1 := \overset{\circ}{H}{}^1(\mathbb{R}^3)$ be the completion of the space $C_0^\infty(\mathbb{R}^3)$ in the norm $\|\nabla \psi(x)\|$.

Definition 1.5.1 (i) $\mathcal{E} := \overset{\circ}{H}{}^1 \oplus L^2 \oplus \mathbb{R}^3 \oplus \mathbb{R}^3$ is the Hilbert phase space of tetrads (ψ, π, q, p) with finite norm

$$\|(\psi, \pi, q, p)\|_{\mathcal{E}} = \|\nabla \psi\| + \|\pi\| + |q| + |p|.$$

(ii) $\mathcal{E}_\sigma$ for $\sigma \in \mathbb{R}$ is the space of $Y = (\psi, \pi, q, p)$ with $\psi \in C^2(\mathbb{R}^3)$ and $\pi \in C^1(\mathbb{R}^3)$, satisfying the estimate

$$|\nabla \psi(x)| + |\pi(x)| + |x|(|\nabla \nabla \psi(x)| + |\nabla \pi(x)|) = \mathcal{O}(|x|^{-\sigma}), \quad |x| \to \infty. \qquad (1.5.5)$$

(iii) $\mathcal{E}_F$ is the space $\mathcal{E}$ with metric of type (1.2.9), where the corresponding seminorms are defined as

$$\|(\psi, \pi, q, p)\|_{\mathcal{E}, R} = \|\nabla \psi\|_R + \|\psi\|_R + \|\pi\|_R + |q| + |p|. \qquad (1.5.6)$$

Obviously, the energy (1.5.4) is a continuous functional on $\mathcal{E}$, and $\mathcal{E}_\sigma \subset \mathcal{E}$ for $\sigma > 3/2$. The convergence in $\mathcal{E}_F$ is equivalent to the convergence in every seminorm (1.5.6). We assume the external potential to be confining:

$$V(q) \to \infty, \qquad |q| \to \infty. \qquad (1.5.7)$$

In this case the Hamiltonian (1.5.4) is bounded below:

$$\inf_{Y \in \mathcal{E}} \mathcal{H}(Y) = V_0 + \frac{1}{2}(\rho, \Delta^{-1}\rho), \qquad (1.5.8)$$

where

$$V_0 := \inf_{q \in \mathbb{R}^3} V(q) > -\infty. \tag{1.5.9}$$

The following lemma is proved in [44, Lemma 2.1].

Lemma 1.5.2 *Let $V(q) \in C^2(\mathbb{R}^3)$ satisfy the condition (1.5.9). Then, for any initial state $Y(0) \in \mathcal{E}$, there exists a unique finite-energy solution $Y(t) = (\psi(t), \pi(t), q(t), p(t)) \in C(\mathbb{R}, \mathcal{E})$, and*
(i) for every $t \in \mathbb{R}$ the map $W(t) : Y_0 \mapsto Y(t)$ is continuous both on $\mathcal{E}$ and on $\mathcal{E}_F$;
(ii) the energy $\mathcal{H}(Y(t))$ is conserved, i.e.,

$$\mathcal{H}(Y(t)) = \mathcal{H}(Y_0) \quad for \ t \in \mathbb{R}; \tag{1.5.10}$$

(iii) a priori estimates hold:

$$\sup_{t \in \mathbb{R}}[\|\nabla \psi(t)\| + \|\pi(t)\|] < \infty, \qquad \sup_{t \in \mathbb{R}} |\dot{q}(t)| = \bar{v} < 1; \tag{1.5.11}$$

(iv) if (1.5.7) holds, then also

$$\sup_{t \in \mathbb{R}} |q(t)| = \bar{q}_0 < \infty. \tag{1.5.12}$$

Remark 1.5.3 In the case of point particle $\rho(x) = \delta(x)$, the system (1.5.1)–(1.5.2) is incorrect, since in this case, any solution of the wave equation (1.5.1) is singular at the point $x = q(t)$, and, accordingly, the integral in (1.5.2) is not defined. Energy functional (1.5.4) in this case is not bounded from below, because the last term in (1.5.8) equals $-\infty$. Indeed, in the Fourier transform, this term has the form

$$(\rho, \Delta^{-1}\rho) = -\int \frac{|\hat{\rho}(k)|^2}{k^2} dk,$$

where $\hat{\rho}(k) \equiv 1$. This is the famous "ultraviolet divergence." Thus, the self-energy of point charge is infinite, which prompted Abraham to introduce the model of an "extended electron" with a continuous charge density $\rho(x)$ [203, 204].

Denote $Z = \{q \in \mathbb{R}^3 : \nabla V(q) = 0\}$. It is easy to verify that stationary states of the system (1.5.1)–(1.5.2) have the form $S_q = (\psi_q, 0, q, 0)$, where $q \in Z$ and $\Delta \psi_q(x) = \rho(x - q)$. Therefore, $\psi_q(x)$ is the Coulomb potential

$$\psi_q(x) := -\frac{1}{4\pi} \int \frac{\rho(y - q) \, dy}{|x - y|}.$$

Respectively, the set of all stationary states of this system is

$$\mathcal{S} := \{S_q : q \in Z\}.$$

If the set Z is discrete in $\mathbb{R}^3$, then the set $\mathcal{S}$ is also discrete in $\mathcal{E}$ and in $\mathcal{E}_F$. Finally, assume that the "form-factor" ρ satisfies the *Wiener condition*

$$\hat{\rho}(k) := \int e^{ikx} \rho(x)\, dx \neq 0, \qquad k \in \mathbb{R}^3. \qquad (1.5.13)$$

Remark 1.5.4 The Wiener condition means a strong coupling of scalar wave field $\psi(x)$ to the particle. It is a suitable version of the "Fermi Golden Rule" for the system (1.5.1)–(1.5.2): the perturbation $\rho(x - q)$ is not orthogonal to all eigenfunctions of continuous spectrum of the Laplacian Δ.

For simplicity of the exposition we assume that

$$\rho \in C_0^\infty(\mathbb{R}^3), \qquad \rho(x) = 0 \text{ for } |x| \geq R_\rho, \qquad \rho(x) = \rho_r(|x|). \qquad (1.5.14)$$

The main result of [44] is as follows.

Theorem 1.5.5 *(i) Let the conditions (1.5.7) and (1.5.13) hold, and $\sigma > 3/2$. Then for any initial state $Y(0) = (\psi_0, \pi_0, q_0, p_0) \in \mathcal{E}_\sigma$, the corresponding solution $Y(t) = (\psi(t), \pi(t), q(t), p(t)) \in C(\mathbb{R}, \mathcal{E})$ to the system (1.5.1)–(1.5.2) attracts to the set of stationary states:*

$$Y(t) \xrightarrow{\mathcal{E}_F} \mathcal{S}, \qquad t \to \pm\infty, \qquad (1.5.15)$$

where attraction holds in the metric (1.2.9) defined with the seminorms (1.5.6).

(ii) Let, additionally, the set Z be discrete in $\mathbb{R}^3$. Then

$$Y(t) \xrightarrow{\mathcal{E}_F} S_\pm \in \mathcal{S}, \qquad t \to \pm\infty. \qquad (1.5.16)$$

The key point in the proof of this theorem is the relaxation of the acceleration

$$\ddot{q}(t) \to 0, \qquad t \to \pm\infty. \qquad (1.5.17)$$

This relaxation has long been known in classical electrodynamics as "radiation damping." Namely, the Liénard–Wiechert formulas for retarded potentials suggest that a particle with a nonzero acceleration radiates energy to infinity. This radiation cannot last forever, because the total energy of the solution is finite. These arguments result in the conclusion (1.5.17), which can be found in any textbook on classical electrodynamics.

However, rigorous proof is not so obvious, and it was done for the first time in [44]. The proof relies on calculation of total energy amount radiated to infinity using the Liénard–Wiechert formulas. The central point is the representation of this amount in the form of a convolution and subsequent application of the Wiener Tauberian theorem.

Below we give a streamlined version of this proof.

Remark 1.5.6 (i) The condition (1.5.7) is not necessary for relaxation (1.5.17). The relaxation also takes place under the condition (1.5.9) (see Remark 1.5.9).

(ii) The Wiener condition (1.5.13) also is not necessary for relaxation (1.5.17). For example, (1.5.17) obviously holds in the case when $V(x) \equiv 0$ and $\rho(x) \equiv 0$. More generally, such relaxation also holds when $V(x) \equiv 0$ and the norm $\|\rho\|$ is sufficiently small (see (2.2.1)).

1.5.1 Liénard–Wiechert Asymptotics

Let us recall long-range asymptotics of the Liénard–Wiechert potentials established in [44, 45]. Denote by $\psi_r(x,t)$ the retarded potential

$$\psi_r(x,t) = -\frac{1}{4\pi} \int \frac{d^3y\, \theta(t - |x-y|)}{|x-y|} \rho(y - q(t - |x-y|)) \quad (1.5.18)$$

and set $\pi_r(x,t) = \dot{\psi}_r(x,t)$. Denote $T_r := \overline{q}_0 + R_\rho$.

Lemma 1.5.7 *The following asymptotics hold:*

$$\left\{ \begin{array}{l} \pi_r(x, |x| + t) = \overline{\pi}(\omega(x),t)|x|^{-1} + \mathcal{O}(|x|^{-2}) \\ \nabla \psi_r(x, |x| + t) = -\omega(x)\overline{\pi}(\omega(x),t)|x|^{-1} + \mathcal{O}(|x|^{-2}) \end{array} \right| , \quad |x| \to \infty$$

$$(1.5.19)$$

uniformly in $t \in [T_r, T]$ for any $T > T_r$. Here $\omega(x) = x/|x|$, and $\overline{\pi}(\omega(x),t)$ is given in (1.5.21).

Proof The integrand of (1.5.18) vanishes for $|y| > T_r$. Then $|x - y| \le t$ for $t - |x| > T_r$, and (1.5.18) implies

$$\nabla \psi_r(x,t) = \int \frac{d^3y}{4\pi |x-y|} n \nabla \rho(y - q(t - |x-y|))$$
$$\cdot \dot{q}(t - |x-y|) + \mathcal{O}(|x|^{-2})$$
$$= -\omega(x)\pi_r(x,t) + \mathcal{O}(|x|^{-2}), \qquad t - |x| > T_r,$$

because $n = \frac{x-y}{|x-y|} = \omega(x) + \mathcal{O}(|x|^{-1})$ for bounded $|y|$. Hence, it suffices to prove asymptotics (1.5.19) for π_r only. We have

$$\pi_r(x,t) = -\int d^3y\, \frac{1}{4\pi |x-y|} \nabla \rho(y - q(\tau)) \cdot \dot{q}(\tau), \quad \tau := t - |x-y|.$$

$$(1.5.20)$$

Replacing t by $|x| + t$ in defining τ, we obtain

$$\tau = |x| + t - |x - y| = t + \omega(x) \cdot y + \mathcal{O}(|x|^{-1})$$
$$= \bar{\tau} + \mathcal{O}(|x|^{-1}), \qquad \bar{\tau} = t + \omega \cdot y,$$

since

$$|x| - |x - y| = |x| - \sqrt{|x|^2 - 2x \cdot y + |y|^2} \sim |x| \left(\frac{x \cdot y}{|x|^2} - \frac{|y|^2}{2|x|^2} \right)$$
$$= \omega(x) \cdot y + \mathcal{O}(|x|^{-1}).$$

Hence, (1.5.20) implies (1.5.19) with

$$\bar{\pi}(\omega, t) := -\frac{1}{4\pi} \int d^3 y \, \nabla \rho(y - q(\bar{\tau})) \cdot \dot{q}(\bar{\tau}). \qquad (1.5.21)$$

$\square$

1.5.2 Free Wave Equation

Consider now the solution $\psi_K(x, t)$ of a free wave equation with initial conditions

$$\psi_K(x, 0) = \psi_0(x), \quad \dot{\psi}_K(x, 0) = \pi_0(x), \qquad x \in \mathbb{R}^3. \qquad (1.5.22)$$

The Kirchhoff formula gives

$$\psi_K(x, t) = \frac{1}{4\pi t} \int_{S_t(x)} d^2 y \, \pi_0(y) + \frac{\partial}{\partial t} \left(\frac{1}{4\pi t} \int_{S_t(x)} d^2 y \, \psi_0(y) \right). \qquad (1.5.23)$$

Here $S_t(x)$ is the sphere $\{y : |y - x| = t\}$. Denote $\pi_K(x, t) = \dot{\psi}_K(x, t)$.

Lemma 1.5.8 *Let $Y_0 \in \mathcal{E}_\sigma$. Then, for any $R > 0$ and any $T_2 > T_1 \geq 0$,*

$$\int_{R+T_1}^{R+T_2} dt \int_{\partial B_R} d^2 x \left(|\pi_K(x, t)|^2 + |\nabla \psi_K(x, t)|^2 \right) \leq I_0 < \infty. \qquad (1.5.24)$$

Proof Formula (1.5.23) implies

$$\nabla \psi_K(x, t) = \frac{t}{4\pi} \int_{S_1} d^2 z \, \nabla \pi_0(x + tz) + \frac{1}{4\pi} \int_{S_1} d^2 z \, \nabla \psi_0(x + tz)$$
$$+ \frac{t}{4\pi} \int_{S_1} d^2 z \, \nabla_x (\nabla \psi_0(x + tz) \cdot z).$$

Here $S_1 := S_1(0)$. From (1.5.5) it follows that

$$|\nabla \psi_K(x,t)| \le C \sum_{s=0}^{1} t^s \int_{S_1} d^2z \, |x + tz|^{-\sigma-1-s}$$

$$= C \sum_{s=0}^{1} \frac{2\pi t^{s-1}}{(\sigma+s-1)|x|} \Big((t-|x|)^{-\sigma-s+1} - (t+|x|)^{-\sigma-s+1} \Big).$$

Therefore,

$$\int_{R+T_1}^{R+T_2} dt \int_{\partial B_R} d^2x \, |\nabla \psi_k(x,t)|^2$$

$$\le C \int_{R+T_1}^{R+T_2} \left[\frac{(t+R)^{2-2\sigma} + (t-R)^{2-2\sigma}}{t^2} + (t-R)^{-2\sigma} \right] dt$$

$$\le C_1 \int_{R+T_1}^{R+T_2} dt \left[\left(1 + \frac{R}{t} \right)^2 + \left(1 - \frac{R}{t} \right)^2 + 1 \right] (t-R)^{-2\sigma} \le I_0 < \infty.$$

The integral with $\nabla \pi_K(x,t)$ can be estimated similarly. $\qquad\square$

1.5.3 Scattering of Energy to Infinity

Now we obtain a bound on the total energy radiated to infinity, which we will represent as a "radiation integral."

This integral has to be bounded a priori by (1.5.11). Indeed, the energy $\mathcal{H}_R(t)$ at time $t \in \mathbb{R}$ in the ball B_R is defined by

$$\mathcal{H}_R(t) = \frac{1}{2} \int_{B_R} d^3x \left(|\pi(x,t)|^2 + |\nabla \psi(x,t)|^2 \right) + \sqrt{1 + p^2(t)} + V(q(t))$$

$$+ \int d^3x \, \psi(x,t)\rho(x - q(t)).$$

Consider the energy $I_R(T_1, T_2)$ radiated from the ball B_R during the time interval $[T_1, T_2]$ with $T_2 > T_1 > 0$:

$$I_R(T_1, T_2) = \mathcal{H}_R(T_1) - \mathcal{H}_R(T_2).$$

This energy is bounded a priori, because by (1.5.11), the energy $\mathcal{H}_R(T_1)$ is bounded from above, while $\mathcal{H}_R(T_2)$ is bounded from below. Thus,

$$I_R(T_1, T_2) \le I < \infty, \tag{1.5.25}$$

where I does not depend on T_1, T_2, and R. Furthermore, one has

$$\frac{d}{dt}\mathcal{H}_R(t) = \int\limits_{\partial B_R} d^2x\, \omega(x) \cdot \pi(x,t)\nabla\psi(x,t), \quad t > R.$$

Hence, (1.5.25) implies

$$\int\limits_{R+T_1}^{R+T_2} dt \int\limits_{\partial B_R} d^2x\, \omega(x) \cdot \pi(x,t)\nabla\psi(x,t) \leq I.$$

The solution admits the splitting $\pi = \pi_r + \pi_K$, $\psi = \psi_r + \psi_K$, and hence,

$$\int\limits_{R+T_1}^{R+T_2} dt \int\limits_{\partial B_R} d^2x\, \omega(x) \cdot (\pi_r\nabla\psi_r + \pi_K\nabla\psi_r + \pi_r\nabla\psi_K + \pi_K\nabla\psi_K) \leq I.$$

Lemmas 1.5.7 and 1.5.8 together with the Cauchy–Schwarz inequality imply

$$\int\limits_{T_r}^{T} dt \int\limits_{S_1} d^2\omega\, |\overline{\pi}(\omega,t)|^2 \leq I_1 + T\mathcal{O}(R^{-1}), \quad T > T_r,$$

where $I_1 < \infty$ does not depend on T and R. Taking the limit $R \to \infty$ and then $T \to \infty$, we obtain the finiteness of the energy radiated to infinity:

$$\int\limits_{0}^{\infty} dt \int\limits_{S_1} d^2\omega |\overline{\pi}(\omega,t)|^2 < \infty. \tag{1.5.26}$$

1.5.4 Convolution Representation and Relaxation of Acceleration

Applying a partial integration in (1.5.21), we obtain

$$\overline{\pi}(\omega,t) = \int d^3y\, \nabla\rho(y - q(\overline{\tau})) \cdot \dot{q}(\overline{\tau})$$

$$= \int d^3y\, \nabla_y\rho(y - q(\overline{\tau})) \cdot \dot{q}(\overline{\tau})\frac{1}{1 - \omega \cdot \dot{q}(\overline{\tau})}$$

$$= -\int d^3y\, \rho(y - q(\overline{\tau}))\frac{\partial}{\partial y_\alpha}\frac{\dot{q}_\alpha(\overline{\tau})}{1 - \omega \cdot \dot{q}(\overline{\tau})}$$

$$= \frac{1}{4\pi}\int d^3y\, \rho(y - q(\overline{\tau}))\frac{\omega \cdot \ddot{q}(\tau)}{(1 - \omega \cdot \dot{q}(\tau))^2}. \tag{1.5.27}$$

The function $\overline{\pi}(\omega,t)$ is globally Lipschitz-continuous in ω and t due to (1.5.11). Hence, (1.5.26) implies

$$\lim_{t\to\infty} \overline{\pi}(\omega,t) = 0 \tag{1.5.28}$$

uniformly in $\omega \in S_1$. Denote $r(t) = \omega \cdot q(t)$, $s = \omega \cdot y$, $\tilde{\rho}(q_3) = \int dq_1 dq_2\, \rho(q_1,q_2,q_3)$ and decompose the y-integration in (1.5.27) along and transversal to ω. Then we obtain the convolution

$$\begin{aligned}
\overline{\pi}(\omega,t) &= \int ds\, \tilde{\rho}(s - r(t+s))\, \frac{\ddot{r}(t+s)}{(1 - \dot{r}(t+s))^2} \\
&= \int d\tau\, \tilde{\rho}(t - (\tau - r(\tau)))\, \frac{\ddot{r}(\tau)}{(1 - \dot{r}(\tau))^2} \\
&= \int d\theta\, \tilde{\rho}(t - \theta) g_\omega(\theta) = \tilde{\rho} * g_\omega(t).
\end{aligned}$$

Here $\theta = \theta(\tau) = \tau - r(\tau)$ is a nondegenerate diffeomorphism of $\mathbb{R}$ since $\dot{r} \le \overline{r} < 1$ due to (1.5.11), and

$$g_\omega(\theta) = \frac{\ddot{r}(\tau(\theta))}{(1 - \dot{r}(\tau(\theta)))^3}. \tag{1.5.29}$$

Let us extend $q(t) = 0$ for $t < 0$. Then $\tilde{\rho} * g_\omega(t)$ is defined for all t and coincides with $\overline{\pi}(\omega,t)$ for sufficiently large t. Hence, (1.5.28) reads as a convolution limit

$$\lim_{t\to\infty} \tilde{\rho} * g_\omega(t) = 0. \tag{1.5.30}$$

Moreover, $g'_\omega(\theta)$ is bounded by (1.5.11). Therefore, (1.5.30) and the Wiener condition (1.5.13) imply

$$\lim_{\theta\to\infty} g_\omega(\theta) = 0, \qquad \omega \in S_1 \tag{1.5.31}$$

by Pitt's extension of the Wiener Tauberian theorem (cf. [15, Thm. 9.7(b)]). Hence, (1.5.29) implies

$$\lim_{t\to\infty} \ddot{q}(t) = 0, \tag{1.5.32}$$

since $\theta(t) \to \infty$ as $t \to \infty$. Finally,

$$\lim_{t\to\infty} \dot{q}(t) = 0, \tag{1.5.33}$$

since $|q(t)| \le \overline{q}_0$ due to (1.5.11).

Remark 1.5.9 (i) We have used condition (1.5.7) in the proof of (1.5.25). However, (1.5.9) at this point is also sufficient. Hence, the relaxation (1.5.32) holds also under condition (1.5.9).

(ii) For point charge $\rho(x) = \delta(x)$, (1.5.30) implies (1.5.31) directly.

(iii) Condition (1.5.13) is necessary for the implication (1.5.31)$\Rightarrow$(1.5.32). Indeed, if (1.5.13) is violated, then $\hat{\rho}_a(\xi) = 0$ for some $\xi \in \mathbb{R}$, and with the choice $g(\theta) - \exp(i\xi\theta)$ we have $\rho_a * g(t) \equiv 0$, whereas g does not decay to zero.

1.5.5 A Compact Attracting Set

Here we show that the set

$$\mathcal{A} = \{S_q : q \in \mathbb{R}^3, |q| \le \overline{q}_0\} \tag{1.5.34}$$

is an attracting subset. It is compact in $\mathcal{E}_F$, since $\mathcal{A}$ is homeomorphic to a closed ball in $\mathbb{R}^3$.

Lemma 1.5.10 *The following attraction holds:*

$$Y(t) \xrightarrow{\mathcal{E}_F} \mathcal{A}, \quad t \longrightarrow \pm\infty. \tag{1.5.35}$$

Proof We need to check that for every $R > 0$,

$$\text{dist}_R(Y(t), \mathcal{A}) = |p(t)| + \|\pi(t)\|_R$$

$$+ \inf_{S_q \in \mathcal{A}} \left(|q(t) - q| + \|\psi(t) - \psi_q\|_R + \|\nabla(\psi(t) - \psi_q)\|_R \right) \to 0 \tag{1.5.36}$$

as $t \to +\infty$. We estimate each summand separately.

(i) $|p(t)| \to 0$ as $t \to \infty$ by (1.5.32).

(ii) $\inf\limits_{|q| \le \overline{q}_0} |q(t) - q| = 0$ for any $t \in \mathbb{R}$ by (1.5.11).

(iii) (1.5.18) implies for $t > R + T_r$ and $|x| < R$

$$|\pi_r(x,t)| \le C \max_{t-R-T_r \le \tau \le t} |\dot{q}(\tau)| \int\limits_{|y|<T_r} d^3y \, \frac{1}{|x-y|} |\nabla\rho(y - q(t - |x-y|))|.$$

The integral in the RHS is bounded uniformly in $t > R + T_r$ and $x \in B_R$. Hence, $\|\pi_r(t)\|_R \to 0$ as $t \to \infty$ by (1.5.33). Then, also, $\|\pi(t)\|_R \to 0$.

(iv) We can replace q with $q(t)$ in the last line of (1.5.36). Then, for $t > R + T_r$ and $|x| < R$, one has

$$\psi_r(x,t) - \psi_{q(t)}(x)$$

$$= - \int\limits_{|y|<T_r} d^3y \, \frac{1}{4\pi|x-y|} \left(\rho(y - q(t - |x-y|)) - \rho(y - q(t)) \right)$$

by (1.5.18). Moreover, $\rho(y - q(t - |x - y|)) - \rho(y - q(t)) \to 0$ as $t \to \infty$ uniformly in $x \in B_R$ due to (1.5.33). Hence, $\|\psi_r(t) - \psi_{q(t)}\|_R \to 0$ as $t \to \infty$. Then, also, $\|\psi(t) - \psi_{q(t)}\|_R \to 0$. Finally, $\|\nabla(\psi(t) - \psi_{q(t)})\|_R$ can be estimated in a similar way. □

1.5.6 Global Attraction to Stationary States

Now we complete the proof of Theorem 1.5.5.

(i) Let $Y(t) \in C(\mathbb{R}, \mathcal{E})$ be any finite-energy solution to the system (1.5.1)–(1.5.2). If the attraction (1.5.15) does not hold, there is a sequence $t_k \to \infty$ for which

$$\text{dist}(Y(t_k), \mathcal{S}) \geq \delta > 0, \qquad k = 1, 2, \ldots . \tag{1.5.37}$$

Since $\mathcal{A}$ is a compact set in $\mathcal{E}_F$, (1.5.35) implies that

$$Y(t_{k'}) \xrightarrow{\mathcal{E}_F} \overline{Y} \in \mathcal{A}, \qquad k' \to \infty \tag{1.5.38}$$

for some subsequence $k' \to \infty$. It remains to check that $\overline{Y} = S_{q_*} \in \mathcal{S}$ with some $q_* \in Z$, since this contradicts (1.5.37).

First, $\overline{Y} = S_q$ with some $|q| \leq \overline{q}_0$ by the definition (1.5.34). Similarly, by the continuity of the map $W(t)$ in $\mathcal{E}_F$,

$$W(t)Y(t_{k'}) = Y(t_{k'} + t) \xrightarrow{\mathcal{E}_F} W(t)\overline{Y} = S_{Q(t)}, \qquad k' \to \infty, \tag{1.5.39}$$

where $Q(\cdot) \in C^2(\mathbb{R}, \mathcal{E})$, since $W(t)\overline{Y} \in C(\mathbb{R}, \mathcal{E})$ is a solution to the system (1.5.1)–(1.5.2). Finally, for $S_{Q(t)}$ to be a solution to the system (1.5.1)–(1.5.2), there must be $\dot{Q}(t) \equiv 0$. Therefore, $Q(t) \equiv q_* \in Z$ and $\overline{Y} = S_{q_*} \in \mathcal{S}$.

(ii) If the set Z is discrete in $\mathbb{R}^3$, then solitary manifold $\mathcal{S}$ is discrete in $\mathcal{E}_F$. □

1.6 Maxwell–Lorentz Equations: Radiation Damping

In [45], global attraction to stationary states similar to (1.5.15), (1.5.16) was established for the Maxwell–Lorentz equations with charged relativistic particle:

$$\begin{cases} \dot{E}(x,t) = \text{rot } B(x,t) - \dot{q}\rho(x-q), & \dot{B}(x,t) = -\text{rot } E(x,t) \\[2mm] \text{div } E(x,t) = \rho(x-q), & \text{div } B(x,t) = 0, & \dot{q}(t) = \dfrac{p(t)}{\sqrt{1+p^2(t)}} \\[2mm] \dot{p}(t) = \displaystyle\int [E(x,t) + E^{\text{ext}}(x,t) + \dot{q}(t) \wedge (B(x,t) + B^{\text{ext}}(x,t))]\rho(x - q(t))\, dx \end{cases} \tag{1.6.1}$$

Here $\rho(x - q)$ is the particle charge density, $\dot{q}\rho(x - q)$ is the corresponding current density, and $E^{\text{ext}} = -\nabla\phi^{\text{ext}}(x)$ and $B^{\text{ext}} = -\text{rot}\,A^{\text{ext}}(x)$ are external static Maxwell fields. Similarly to (1.5.7), we assume that *effective scalar potential* is confining:

$$V(q) := \int \phi^{\text{ext}}(x)\rho(x - q)\,dx \to \infty, \qquad |q| \to \infty. \tag{1.6.2}$$

This system describes classical electrodynamics with "extended electron" introduced by M. Abraham [203, 204]. In the case of a point electron, when $\rho(x) = \delta(x)$, such a system is not well defined. Indeed, in this case, any solutions $E(x,t)$ and $B(x,t)$ of the Maxwell equations (the first line of (1.6.1)) are singular for $x = q(t)$, and accordingly, the integral in the last equation (1.6.1) does not exist.

This system may be formally presented in Hamiltonian form if the fields are expressed in terms of potentials $E(x,t) = -\nabla\phi(x,t) - \dot{A}(x,t)$, $B(x,t) = -\text{rot}\,A(x,t)$. The corresponding Hamiltonian functional reads

$$\mathcal{H} = \frac{1}{2}[\langle E, E\rangle + \langle B, B\rangle] + V(q) + \sqrt{1 + p^2}$$

$$= \frac{1}{2}\int [E^2(x) + B^2(x)]\,dx + V(q) + \sqrt{1 + p^2}. \tag{1.6.3}$$

The Hilbert phase space of finite-energy states is defined as $\mathcal{E} := L^2 \oplus L^2 \oplus \mathbb{R}^3 \oplus \mathbb{R}^3$. Under the condition (1.6.2) a solution $Y(t) = (E(x,t), B(x,t), q(t), p(t)) \in C(\mathbb{R}, \mathcal{E})$ of finite energy exists and is unique for any initial state $Y(0) \in \mathcal{E}$.

The Hamiltonian (1.6.3) is conserved along solutions, and it provides *a priori estimates*, which play an important role in proving global attraction of the type (1.5.15), (1.5.16) in [45]. The key role in the proof is played again by relaxation of acceleration (1.5.17), which is derived by a suitable generalization of our methods [44]: the expression of energy radiated to infinity via Liénard–Wiechert retarded potentials, its representation in the form of a convolution, and the use of the Wiener Tauberian theorem.

In classical electrodynamics the **radiation damping** (1.5.17) is traditionally derived from the Larmor and Liénard formulas for radiation power of a point particle (see formulas (14.22) and (14.24) of [212]), but this approach ignores field feedback, although it plays the key role in the relaxation of the acceleration. The main problem is that this reverse field reaction for point particles is infinite. A rigorous sense of these classical calculations was first found in [44, 45] for the Abraham model of "extended electron" under the Wiener condition (1.5.13). A detailed discussion can be found in [52].

1.7 Wave Equations with Concentrated Nonlinearities

Here we prove the result of [50] on global attraction to stationary states for a 3D wave equation with point coupling to a $U(1)$-invariant nonlinear oscillator. This goal is inspired by the fundamental mathematical problem of an interaction of point particles with the fields.

Point interaction models have been considered since 1930 in the papers of E. Wigner, H. Bethe and R. Peierls, E. Fermi, and others (see [90] for a detailed survey) and of Dirac [93]. Rigorous mathematical results have been obtained since 1960 by Ya. B. Zeldovich, F. Berezin, L. Faddeev, F. H. J. Cornish, D. Yafaev, E. Zeidler, and others [91, 92, 94, 96, 98], and since 2000 by D. Noja, A. Posilicano, and others [95, 97, 89].

We consider a real wave field $\psi(x,t)$ coupled to a nonlinear oscillator:

$$
\begin{cases}
\ddot{\psi}(x,t) = \Delta\psi(x,t) + \zeta(t)\delta(x) \\
\lim_{x \to 0}(\psi(x,t) - \zeta(t)G(x)) = F(\zeta(t))
\end{cases}
\quad\quad x \in \mathbb{R}^3, \quad t \in \mathbb{R}, \quad (1.7.1)
$$

where $G(x) = \frac{1}{4\pi|x|}$ is the Green function of the operator $-\Delta$ in $\mathbb{R}^3$. Nonlinear function $F(\zeta)$ admits a potential:

$$
F(\zeta) = U'(\zeta), \quad \zeta \in \mathbb{R}, \quad U \in C^2(\mathbb{R}). \quad\quad (1.7.2)
$$

We assume that the potential is confining, i.e.,

$$
U(\zeta) \to \infty, \quad \zeta \to \pm\infty. \quad\quad (1.7.3)
$$

The system (1.7.1) admits stationary solutions $\psi_q = qG(x) \in L^2_{\text{loc}}(\mathbb{R}^3)$, where $q \in Q := \{q \in \mathbb{R} : F(q) = 0\}$. We assume that the set Q is nonempty and does not contain intervals, i.e.,

$$
[a,b] \not\subset Q \quad\quad (1.7.4)
$$

for any $a < b$.

As before, $\|\cdot\|$ and $\|\cdot\|_R$ denote the norms in $L^2 = L^2(\mathbb{R}^3)$ and in $L^2(B_R)$, respectively, and $\overset{\circ}{H}{}^1 = \overset{\circ}{H}{}^1(\mathbb{R}^3)$ is the completion of the space $C_0^\infty(\mathbb{R}^3)$ in the norm $\|\nabla\psi(x)\|$. Denote

$$
\overset{\circ}{H}{}^2 = \overset{\circ}{H}{}^2(\mathbb{R}^3) := \{f \in \overset{\circ}{H}{}^1, \ \Delta f \in L^2\}, \quad t \in \mathbb{R}.
$$

We define the function sets

$$
D = \{\psi \in L^2 : \psi(x) = \psi_{\text{reg}}(x) + \zeta G(x),
$$
$$
\psi_{\text{reg}} \in \overset{\circ}{H}{}^2, \ \zeta \in \mathbb{R}, \ \lim_{x \to 0}\psi_{\text{reg}}(x) = F(\zeta)\}
$$

and

$$\dot{D} = \{\pi \in L^2(\mathbb{R}^3) : \pi(x) = \pi_{\text{reg}}(x) + \eta G(x), \ \pi_{\text{reg}} \subset \overset{\circ}{H}{}^1, \ \eta \in \mathbb{R}\}.$$

Obviously, $D \subset \dot{D}$.

Definition 1.7.1 $\mathcal{D}$ is the Hilbert manifold of states $\Psi = (\psi, \pi) \in D \times \dot{D}$.

First, we prove global well-posedness for the system (1.7.1) established in [95].

Theorem 1.7.2 *Let conditions (1.7.2) and (1.7.3) hold. Then*
(i) For all initial data $\Psi_0 = (\psi_0, \pi_0) \in \mathcal{D}$, the system (1.7.1) has a unique solution $\Psi(t) = (\psi(t), \dot{\psi}(t)) \in C(\mathbb{R}, \mathcal{D})$.
(ii) The energy is conserved:

$$\mathcal{H}(\Psi(t)) := \frac{1}{2}\Big(\|\dot{\psi}(t)\|^2 + \|\nabla\psi_{\text{reg}}(t)\|^2\Big) + U(\zeta(t)) = \text{const}, \quad t \in \mathbb{R}.$$
(1.7.5)

(iii) The following a priori bound holds:

$$|\zeta(t)| \le C(\Psi_0), \quad t \in \mathbb{R}.$$
(1.7.6)

Proof It suffices to prove the theorem for $t \ge 0$.

Step (i). First we consider the free wave equation with initial data from $\mathcal{D}$:

$$\ddot{\psi}_f(x,t) = \Delta\psi_f(x,t), \quad (\psi_f(0), \dot{\psi}_f(0)) = (\psi_0, \pi_0)$$
$$= (\psi_{0,\text{reg}}, \pi_{0,\text{reg}}) + (\zeta_0 G, \eta_0 G) \in \mathcal{D},$$
(1.7.7)

where $(\psi_{0,\text{reg}}, \pi_{0,\text{reg}}) \in \overset{\circ}{H}{}^2 \oplus \overset{\circ}{H}{}^1$.

Lemma 1.7.3 *There exists a unique solution $\psi_f(t) \in C([0;\infty), L^2_{\text{loc}}(\mathbb{R}^3))$ to (1.7.7). Moreover, for any $t > 0$, there exists the limit*

$$\lambda(t) := \lim_{x \to 0} \psi_f(x,t) \in C[0,\infty),$$

and

$$\dot{\lambda}(t) \in L^2_{\text{loc}}[0,\infty).$$
(1.7.8)

Proof We split $\psi_f(x,t)$ as

$$\psi_f(x,t) = \psi_{f,\text{reg}}(x,t) + g(x,t),$$

where $\psi_{f,\text{reg}}$ and g are solutions to the free wave equation with initial data $(\psi_{0,\text{reg}}, \pi_{0,\text{reg}})$ and $(\zeta_0 G, \eta_0 G)$, respectively. First, $\psi_{f,\text{reg}} \in C([0,\infty), \overset{\circ}{H}{}^2)$

by the energy conservation. Hence, $\lim_{x \to 0} \psi_{f,\mathrm{reg}}(x,t)$ exists for any $t \geq 0$, since $\overset{\circ}{H}{}^2(\mathbb{R}^3) \subset C(\mathbb{R}^3)$.

Let us obtain an explicit formula for g. Note that the function $h(x,t) = g(x,t) - (\zeta_0 + \eta_0 t)G(x)$ satisfies

$$\ddot{h}(x,t) = \Delta h(x,t) - (\zeta_0 + \eta_0 t)\delta(x), \quad h(x,0) = 0, \; \dot{h}(x,0) = 0. \quad (1.7.9)$$

The unique solution to (1.7.9) is a spherical wave:

$$h(x,t) = -\frac{\theta(t - |x|)}{4\pi|x|}(\zeta_0 + \eta_0(t - |x|)), \quad t \geq 0. \quad (1.7.10)$$

Here θ is the Heaviside function. Hence,

$$g(x,t) = h(x,t) + (\zeta_0 + \eta_0 t)G(x)$$
$$= -\frac{\theta(t - |x|)(\zeta_0 + \eta_0(t - |x|))}{4\pi|x|} + \frac{\zeta_0 + \eta_0 t}{4\pi|x|} \in C([0,\infty), L^2_{\mathrm{loc}}(\mathbb{R}^3)),$$

and then

$$\lim_{x \to 0} g(x,t) = \frac{\eta_0}{4\pi}, \quad t > 0.$$

Finally, $\dot{\psi}_{f,\mathrm{reg}}(0,t) \in L^2_{\mathrm{loc}}([0,\infty))$ by [50, Lemma 3.4]. Hence, (1.7.8) follows. $\qquad \square$

Step (ii). Now we prove local well-posedness. We modify the nonlinearity F so that it becomes Lipschitz-continuous. Define

$$\Lambda(\Psi_0) = \sup\{|\zeta| : \zeta \in \mathbb{R}, \, U(\zeta) \leq \mathcal{H}(\Psi_0)\}.$$

We may pick a modified potential function $\tilde{U}(\zeta) \in C^2(\mathbb{R})$, so that

$$\begin{cases} \tilde{U}(\zeta) = U(\zeta), & |\zeta| \leq \Lambda(\Psi_0), \\ \tilde{U}(\zeta) > \mathcal{H}(\Psi_0), & |\zeta| > \Lambda(\Psi_0), \end{cases} \quad (1.7.11)$$

and the function $\tilde{F}(\zeta) = \tilde{U}'(\zeta)$ is Lipschitz-continuous:

$$|\tilde{F}(\zeta_1) - \tilde{F}(\zeta_2)| \leq C|\zeta_1 - \zeta_2|, \quad \zeta_1, \zeta_2 \in \mathbb{R}.$$

The following lemma is trivial.

Lemma 1.7.4 *For small $\tau > 0$ the Cauchy problem*

$$\frac{1}{4\pi}\dot{\zeta}(t) + \tilde{F}(\zeta(t)) = \lambda(t), \quad \zeta(0) = \zeta_0 \quad (1.7.12)$$

has a unique solution $\zeta \in C^1[0,\tau]$.

Denote

$$\psi_S(t,x) := \frac{\theta(t-|x|)}{4\pi|x|}\zeta(t-|x|), \quad t \in [0,\tau],$$

with ζ from Lemma 1.7.4.

Lemma 1.7.5 *The function $\psi(x,t) := \psi_f(x,t) + \psi_S(x,t)$ is a unique solution to the system*

$$\begin{cases} \ddot{\psi}(x,t) = \Delta\psi(x,t) + \zeta(t)\delta(x) \\ \lim_{x\to 0}(\psi(x,t) - \zeta(t)G(x)) = \tilde{F}(\zeta(t)) \\ \psi(x,0) = \psi_0(x), \quad \dot{\psi}(x,0) = \pi_0(x) \end{cases} \quad x \in \mathbb{R}^3, \quad t \in [0,\tau], \quad (1.7.13)$$

satisfying the condition

$$(\psi(t), \dot{\psi}(t)) \in \mathcal{D}, \quad t \in [0,\tau]. \quad (1.7.14)$$

Proof Initial conditions of (1.7.13) follow from (1.7.7). Furthermore,

$$\lim_{x\to 0}(\psi(t,x) - \zeta(t)G(x)) = \lambda(t) + \lim_{x\to 0}\left(\frac{\theta(t-|x|)\zeta(t-|x|)}{4\pi|x|} - \frac{\zeta(t)}{4\pi|x|}\right)$$

$$= \lambda(t) - \frac{1}{4\pi}\dot{\zeta}(t) = \tilde{F}(\zeta(t)).$$

Thus, the second equation of (1.7.13) is satisfied. At last,

$$\ddot{\psi} = \ddot{\psi}_f + \ddot{\psi}_S = \Delta\psi_f + \Delta\psi_S + \zeta\delta = \Delta\psi + \zeta\delta,$$

and ψ solves the first equation of (1.7.13) then.

It remains to check (1.7.14). Note that the function $\varphi_{\mathrm{reg}}(x,t) = \psi(x,t) - \zeta(t)G_1(x) = \psi_{\mathrm{reg}}(x,t) + \zeta(t)(G(x) - G_1(x))$, where $G_1(x) = G(x)e^{-|x|}$, satisfies

$$\ddot{\varphi}_{\mathrm{reg}}(x,t) = \Delta\varphi_{\mathrm{reg}}(x,t) + (\zeta(t) - \ddot{\zeta}(t))G_1(x)$$

with initial data from $H^2 \oplus H^1$. Moreover, (1.7.8) and (1.7.12) imply that $\ddot{\zeta} \in L^2([0,\tau])$. Hence,

$$(\varphi_{\mathrm{reg}}(x,t), \dot{\varphi}_{\mathrm{reg}}(x,t)) \in H^2 \oplus H^1, \quad t \in [0,\tau]$$

by [50, Lemma 3.2]. Therefore,

$$\psi_{\mathrm{reg}}(x,t) = \psi(x,t) - \zeta(t)G(x) = \varphi_{\mathrm{reg}}(x,t) + \zeta(t)(G_1(x) - G(x))$$

satisfies $(\psi_{\mathrm{reg}}(t), \dot{\psi}_{\mathrm{reg}}(t)) \in \mathring{H}^2 \oplus \mathring{H}^1$, $t \in [0,\tau]$, and (1.7.14) holds then.

It remains to prove the uniqueness. Suppose now that there exists another solution $\tilde{\psi} = \tilde{\psi}_{\mathrm{reg}} + \tilde{\zeta}G$ to the system (1.7.13), with $(\tilde{\psi}, \dot{\tilde{\psi}}) \in \mathcal{D}$. Then, by reversing the above argument, the second equation of (1.7.13) implies that $\tilde{\zeta}$ solves the Cauchy problem (1.7.12). The uniqueness of the solution of (1.7.12) implies that $\tilde{\zeta} = \zeta$. Then, defining

$$\psi_S(t,x) := \frac{\theta(t - |x|)}{4\pi |x|} \zeta(t - |x|), \quad t \in [0, \tau]$$

for $\tilde{\psi}_f = \tilde{\psi} - \psi_S$, one obtains

$$\ddot{\tilde{\psi}}_f = \ddot{\tilde{\psi}} - \ddot{\psi}_S = \Delta\tilde{\psi}_{\text{reg}} - (\Delta\psi_S + \zeta\delta) = \Delta(\tilde{\psi}_{\text{reg}} - (\psi_S - \zeta G)) = \Delta\tilde{\psi}_f;$$

i.e., $\tilde{\psi}_f$ solves the Cauchy problem (1.7.7). Hence, $\tilde{\psi}_f = \psi_f$ by the uniqueness of the solution to (1.7.7), and hence, $\tilde{\psi} = \psi$.   □

According to [50, Lemma 3.7]

$$\mathcal{H}_{\tilde{F}}(\Psi(t)) = \|\dot{\psi}(t)\|^2 + \|\nabla\tilde{\psi}_{\text{reg}}(t)\|^2 + \tilde{U}(\zeta(t)) = \text{const}, \quad t \in [0, \tau]. \tag{1.7.15}$$

Now we are able to prove Theorem 1.7.2 on the global well-posedness. First, note that

$$\tilde{U}(\zeta(t)) = U(\zeta(t)), \quad t \in [0, \tau]. \tag{1.7.16}$$

Indeed, $\mathcal{H}_F(\Psi_0) \geq U(\zeta_0)$ by the definition of energy in (1.7.5). Therefore, $|\zeta_0| \leq \Lambda(\Psi_0)$, and then $\tilde{U}(\zeta_0) = U(\zeta_0)$, $\mathcal{H}_{\tilde{F}}(\Psi_0) = \mathcal{H}_F(\Psi_0)$. Furthermore,

$$\mathcal{H}_F(\Psi_0) = \mathcal{H}_{\tilde{F}}(\Psi(t)) \geq \tilde{U}(\zeta(t)), \quad t \in [0, \tau],$$

and (1.7.11) implies that

$$|\zeta(t)| \leq \Lambda(\Psi_0), \quad t \in [0, \tau]. \tag{1.7.17}$$

Now we can replace $\tilde{F}$ by F in Lemma 1.7.5 and in (1.7.15). The solution $\Psi(t) = (\psi(t), \dot{\psi}(t)) \in \mathcal{D}$ constructed in Lemma 1.7.5 exists for $0 \leq t \leq \tau$, where the time span τ in Lemma 1.7.4 depends only on $\Lambda(\Psi_0)$. Hence, the bound (1.7.17) at $t = \tau$ allows us to extend the solution Ψ to the time interval $[\tau, 2\tau]$. We proceed by induction to obtain the solution for all $t \geq 0$. Theorem 1.7.2 is proved.   □

The main result of [50] is as follows.

Theorem 1.7.6 *Let* $\Psi(x,t) = (\psi(x,t), \dot{\psi}(x,t))$ *be a solution to (1.7.1) with initial data from* $\mathcal{D}$. *Then*

$$\Psi(x,t) \to (\psi_{q_\pm}, 0), \quad t \to \pm\infty,$$

where $q_\pm \in Q$ *and the convergence holds in* $L^2_{\text{loc}}(\mathbb{R}^3) \oplus L^2_{\text{loc}}(\mathbb{R}^3)$.

Proof It suffices to prove this theorem for $t \to +\infty$ only. By Lemma 1.7.5, the solution $\psi(x,t)$ to (1.7.1) with initial data $(\psi_0, \pi_0) \in \mathcal{D}$, can be represented as the sum

$$\psi(x,t) := \psi_f(x,t) + \psi_S(x,t), \quad t \geq 0, \tag{1.7.18}$$

where *dispersive component* $\psi_f(x,t)$ is a unique solution to (1.7.7) and *singular component* $\psi_S(x,t)$ is a unique solution to the following Cauchy problem:

$$\ddot{\psi}_S(x,t) = \Delta\psi_S(x,t) + \zeta(t)\delta(x), \quad \psi_S(x,0) = 0, \quad \dot{\psi}_S(x,0) = 0. \tag{1.7.19}$$

Here $\zeta(t) \in C_b^1([0,\infty))$ is a unique solution to

$$\frac{1}{4\pi}\dot{\zeta}(t) + F(\zeta(t)) = \lambda(t), \quad \zeta(0) = \zeta_0. \tag{1.7.20}$$

Now we can prove local decay of $\psi_f(x,t)$.

Lemma 1.7.7 *For any $R > 0$ the following convergence holds:*

$$\left\|(\psi_f(t), \dot{\psi}_f(t))\right\|_{H^2(B_R) \oplus H^1(B_R)} \to 0, \quad t \to \infty. \tag{1.7.21}$$

Here B_R is the ball of radius R.

Proof We represent the initial data $(\psi_0, \pi_0) = (\psi_{0,\text{reg}}, \pi_{0,\text{reg}}) + (\zeta_0 G, \eta_0 G) \in \mathcal{D}$ as

$$(\psi_0, \pi_0) = (\varphi_0, p_0) + (\zeta_0 \chi G, \eta_0 \chi G),$$

where a cutoff function $\chi \in C_0^\infty(\mathbb{R}^3)$ satisfies

$$\chi(x) = \begin{cases} 1, & |x| \leq 1, \\ 0, & |x| \geq 2. \end{cases} \tag{1.7.22}$$

Let us show that

$$(\varphi_0, p_0) \in H^2 \oplus H^1.$$

Indeed,

$$(\varphi_0, p_0) = (\psi_0 - \zeta_0 \chi G, \pi_0 - \eta_0 \chi G) \in L^2 \oplus L^2.$$

On the other hand,

$$(\varphi_0, p_0) = (\psi_{0,\text{reg}} + \zeta_0(1-\chi)G, \pi_{0,\text{reg}} + \eta_0(1-\chi)G) \in \mathring{H}^2 \oplus \mathring{H}^1.$$

Now we split the dispersion component $\psi_f(x,t)$ as

$$\psi_f(x,t) = \varphi(x,t) + \varphi_G(x,t), \quad t \geq 0,$$

where φ and φ_G are defined as solutions to the free wave equation with initial data $(\varphi_0, \ p_0)$ and $(\zeta_0 \chi G, \eta_0 \chi G)$, respectively, and study the decay properties of φ_G and φ.

First, by the strong Huygens principle,

$$\varphi_G(x,t) = 0 \ \text{ for } \ |x| \leq t - 2.$$

Indeed, $\varphi_G(x,t) = \zeta_0 \dot{\psi}_G(x,t) + \eta_0 \psi_G(x,t)$, where $\psi_G(x,t)$ is the solution to the free wave equation with initial data $(0, \chi G) \in H^1 \oplus L^2$ and $\psi_G(x,t)$ satisfies the strong Huygens principle by Theorem XI.87 of [13], v. III.

It remains to check that

$$\|(\varphi(t), \dot{\varphi}(t))\|_{H^2(B_R) \oplus H^1(B_R)} \to 0, \quad t \to \infty, \quad \forall R > 0. \tag{1.7.23}$$

For $r \geq 1$, denote $\chi_r = \chi(x/r)$, where $\chi(x)$ is a cutoff function (1.7.22). Denote $\phi_0 = (\varphi_0, \pi_0)$. Let $u_r(t)$ and $v_r(t)$ be solutions to the free wave equations with the initial data $\chi_r \phi_0$ and $(1 - \chi_r)\phi_0$, respectively, so that $\varphi(t) = u_r(t) + v_r(t)$. By the strong Huygens principle,

$$u_r(x,t) = 0 \ \text{ for } \ t \geq |x| + 2r.$$

To conclude (1.7.23), it remains to note that

$$
\begin{aligned}
\|(v_r(t), \dot{v}_r(t))\|_{H^2(B_R) \oplus H^1(B_R)} &\leq C(R) \|(v_r(t), \dot{v}_r(t))\|_{\overset{\circ}{H}{}^2 \oplus H^1} \\
&= C(R) \|(1 - \chi_r)\phi_0\|_{\overset{\circ}{H}{}^2 \oplus H^1} \\
&\leq C(R) \|(1 - \chi_r)\phi_0\|_{H^2 \oplus H^1} \tag{1.7.24}
\end{aligned}
$$

by the energy conservation for the free wave equation. We also use the Sobolev embedding theorem $\overset{\circ}{H}{}^1(\mathbb{R}^3) \subset L^6(\mathbb{R}^3)$. The RHS of (1.7.24) could be made arbitrarily small if $r \geq 1$ is sufficiently large. $\qquad \square$

Due to (1.7.18) and (1.7.21), for the proof of Theorem 1.7.6 it suffices to verify the convergence of $\psi_S(x,t)$ to stationary states.

Lemma 1.7.8 *Let $\psi_S(x,t)$ and $\zeta(t)$ be solutions to (1.7.19) and (1.7.20), respectively. Then*

$$(\psi_S(t), \dot{\psi}_S(t)) \to (\psi_{q_\pm}, 0), \quad t \to \infty,$$

where $q_\pm \in Q$ and the convergence holds in $L^2_{\text{loc}}(\mathbb{R}^3) \oplus L^2_{\text{loc}}(\mathbb{R}^3)$.

Proof The unique solution to (1.7.19) is the spherical wave

$$\psi_S(x,t) = \frac{\theta(t - |x|)}{4\pi|x|} \zeta(t - |x|), \quad t \geq 0 \tag{1.7.25}$$

(cf. (1.7.9)–(1.7.10)). Then a priori bound (1.7.6) and equation (1.7.20) imply that

$$(\psi_S(t), \dot\psi_S(t)) \in L^2(B_R) \oplus L^2(B_R), \quad 0 \le R < t.$$

First, we prove the convergence of $\zeta(t)$. From (1.7.6) it follows that $\zeta(t)$ has the upper and lower limits

$$\underline{\lim}_{t\to\infty}\zeta(t) = a, \quad \overline{\lim}_{t\to\infty}\zeta(t) = b. \tag{1.7.26}$$

Suppose that $a < b$. Then the trajectory $\zeta(t)$ oscillates between a and b. Assumption (1.7.4) implies that $F(\zeta_0) \ne 0$ for some $\zeta_0 \in (a,b)$. For the concreteness, let us assume that $F(\zeta_0) > 0$. The convergence (1.7.21) implies that

$$\lambda(t) = \psi_f(0,t) \to 0, \quad t \to \infty. \tag{1.7.27}$$

Hence, for sufficiently large T, we have

$$-F(\zeta_0) + \lambda(t) < 0, \quad t \ge T.$$

Then, for $t \ge T$, the transition of the trajectory from left to right through the point ζ_0 is impossible by (1.7.20). Therefore, $a = b = q_+$, where $q_+ \in Q$ since $F(q_+) = 0$ by (1.7.20). Hence (1.7.26) implies

$$\zeta(t) \to q_+, \quad t \to \infty. \tag{1.7.28}$$

Furthermore,

$$\theta(t - |x|) \to 1, \quad t \to \infty \tag{1.7.29}$$

uniformly in $|x| \le R$. Then (1.7.25) and (1.7.28) imply that

$$\psi_S(t) \to q_+ G, \quad t \to \infty,$$

where the convergence holds in $L^2_{\text{loc}}(\mathbb{R}^3)$. It remains to verify the convergence of $\dot\psi_S(t)$. We have

$$\dot\psi_S(x,t) = \frac{\theta(t - |x|)}{4\pi|x|}\dot\zeta(t - |x|), \quad |x| < t.$$

From (1.7.20), (1.7.27), and (1.7.28) it follows that $\dot\zeta(t) \to 0$ as $t \to \infty$. Then

$$\dot\psi_S(t) \to 0, \quad t \to \infty$$

in $L^2_{\text{loc}}(\mathbb{R}^3)$ by (1.7.29). $\qquad\square$

This completes the proof of Theorem 1.7.6. $\qquad\square$

1.8 Comparison with Dissipative Systems

All of the above results on global attraction to stationary states refer to "generic" systems with a trivial symmetry group. These systems are characterized by a suitable discreteness of attractors, by Wiener condition, and so on.

Global attraction to stationary states (7) resembles similar asymptotics (1) for dissipative systems. However, there are a number of fundamental differences:

I. In dissipative systems,

- the (point) global attractor always consists of *stationary states,*
- the global attraction (7) to stationary states is due to the energy absorption,
- the global attraction (1) *holds only as $t \to +\infty$,*
- this attraction can hold *in bounded and unbounded domains,*
- this attraction is due to the absorption of energy and holds mainly in suitable global norms,
- such global attraction to stationary states also *holds for all finite-dimensional dissipative systems.*

II. On the other hand, in Hamiltonian systems,

- the *global attractor may differ from the set of stationary states,* as will be seen below,
- the global attraction (7) to stationary states is due to the *radiation of energy to infinity,* which plays the role of energy absorption,
- this attraction takes place both *as $t \to \infty$ and as $t \to -\infty$,*
- this attraction holds *only in unbounded domains,*
- the attraction holds *only in local seminorms,*
- the attraction to a *proper subset* cannot hold for finite-dimensional Hamiltonian systems due to energy conservation.

2

Global Attraction to Solitons

In this chapter we present the first results [54, 57] on global attraction to solitons (12) for the scalar wave field coupled to a charged relativistic particle with density of charge satisfying the Wiener condition. This result was extended in [55, 56] to a similar system with the Maxwell field.

In particular, the radiation damping in classical electrodynamics is rigorously proved for the first time (in this chapter – for the translation-invariant case). The proofs rely on canonical transformation to the comoving frame, energy bound from below, and on the Wiener Tauberian theorem.

2.1 Translation-Invariant Wave–Particle System

In [54] the system (1.5.1)–(1.5.2) was considered in the case of zero potential $V(x) \equiv 0$:

$$\begin{cases} \ddot{\psi}(x,t) = \Delta \psi(x,t) - \rho(x - q(t)), & x \in \mathbb{R}^3 \\ \dot{q}(t) = \dfrac{p(t)}{\sqrt{1 + p^2(t)}}, \quad \dot{p}(t) = -\displaystyle\int \nabla \psi(x,t) \rho(x - q(t)) \, dx \end{cases}, \quad (2.1.1)$$

which can be written in the Hamiltonian form (1.5.3). The Hamiltonian of this system is given by (1.5.4) with $V = 0$, and it is conserved along trajectories. By Lemma 1.5.2 with $V(x) \equiv 0$, global solutions exist for all initial data $Y(0) \in \mathcal{E}$, and a priori estimates (1.5.11) hold.

This system is translation-invariant, so the corresponding full momentum

$$P = p - \int \pi(x) \nabla \psi(x) \, dx \tag{2.1.2}$$

is also conserved. Respectively, the system (2.1.1) admits traveling-wave-type solutions (solitons):

$$\psi_v(x - a - vt), \quad q(t) = a + vt, \quad p_v = v/\sqrt{1 - v^2}, \qquad (2.1.3)$$

where $v, a \in \mathbb{R}^3$ and $|v| < 1$. The solitons are easily determined: for $|v| < 1$, there is a unique function ψ_v that makes (2.1.3) a solution to (2.1.1),

$$\psi_v(x) = -\int d^3 y (4\pi |(y - x)_\| + \lambda(y - x)_\perp|)^{-1} \rho(y), \qquad (2.1.4)$$

where we set $\lambda = \sqrt{1 - v^2}$ and $x = x_\| + x_\perp$, where $x_\| \| v$ and $x_\perp \perp v$ for $x \in \mathbb{R}^3$. Indeed, substituting (2.1.3) into the wave equation of (2.1.1), we get the stationary equation

$$(v \cdot \nabla)^2 \psi_v(x) = \Delta \psi_v(x) - \rho(x). \qquad (2.1.5)$$

Through the Fourier transform,

$$\hat{\psi}_v(k) = -\hat{\rho}(k)/(k^2 - (v \cdot k)^2), \qquad (2.1.6)$$

which implies (2.1.4). The set of all solitons forms a six-dimensional *solitary manifold* in the Hilbert phase space $\mathcal{E}$:

$$S = \{S_{v,a} = (\psi_v(x - a), \pi_v(x - a), a, p_v) : \quad v, a \in \mathbb{R}^3, \quad |v| < 1\}, \qquad (2.1.7)$$

where $\pi_v := -v \nabla \psi_v$. Recall that the spaces $\mathcal{E}$ and $\mathcal{E}_\sigma$ and the corresponding norms were introduced in Definition 1.5.1. The following theorem is the main result of [54].

Theorem 2.1.1 *Let the Wiener condition (1.5.13) hold and $\sigma > 3/2$. Then, for any initial state $Y(0) \in \mathcal{E}_\sigma$, the corresponding solution $Y(t) = (\psi(t), \pi(t), q(t), p(t))$ of the system (2.1.1) converges to the solitary manifold S in the following sense:*

$$\ddot{q}(t) \to 0, \quad \dot{q}(t) \to v_\pm, \qquad t \to \pm\infty, \qquad (2.1.8)$$

$$(\psi(x,t), \dot{\psi}(x,t)) = (\psi_{v_\pm}(x - q(t)), \pi_{v_\pm}(x - q(t))) + (r_\pm(x,t), s_\pm(x,t)), \qquad (2.1.9)$$

*where the remainder decreases locally in the **comoving frame**: for each $R > 0$,*

$$\|\nabla r_\pm(q(t) + x, t)\|_R + \|r_\pm(q(t) + x, t)\|_R$$
$$+ \|s_\pm(q(t) + x, t)\|_R \to 0, \qquad t \to \pm\infty. \qquad (2.1.10)$$

The theorem means that, in particular,

$$\psi(x,t) \sim \psi_v(x - v_\pm t + \theta_\pm(t)), \quad \text{where} \quad \dot{\theta}_\pm(t) \to 0, \quad t \to \pm\infty.$$
$$(2.1.11)$$

The proof [54] relies on (i) relaxation of acceleration (1.5.17) in the case $V = 0$ (see Remark 1.5.9 (i)) and (ii) the *canonical change of variables* to the comoving frame. The key role is played by the fact that the soliton $S_{v,a}$ minimizes the Hamiltonian (1.5.4) (in the case $V = 0$) with a fixed total momentum (2.1.2), which implies *orbital stability of solitons*; see [100, 101]. In addition, the proof essentially relies on the *strong Huygens principle* for the 3D wave equation.

Before entering into more precise and technical discussion, it may be useful to give a general idea of our strategy. As was mentioned above, the total momentum (2.1.2) is conserved because of translation invariance.

We transform the system (2.1.1) to new variables $(\Psi(x), \Pi(x), Q, P) = (\psi(q+x), \pi(q+x), q, P(\psi, q, \pi, p))$. The key role in our strategy is played by the fact that this transformation is canonical, which is proved in Section 2.1.4. Through this canonical transformation, one obtains the new Hamiltonian

$$\mathcal{H}_P(\Psi, \Pi) = \mathcal{H}(\psi, \pi, q, p)$$
$$= \int d^3x \left(\frac{1}{2}|\Pi(x)|^2 + \frac{1}{2}|\nabla\Psi(x)|^2 + \Psi(x)\rho(x) \right)$$
$$+ \left[1 + \left(P + \int d^3x\, \Pi(x)\, \nabla\Psi(x) \right)^2 \right]^{1/2}.$$

Since Q is the cyclic coordinate (i.e., the Hamiltonian $\mathcal{H}_P$ does not depend on Q), we may regard P as a fixed parameter and consider the reduced system for (Ψ, Π) only. Let us define

$$\pi_v(x) = -v \cdot \nabla\psi_v(x),$$

$$P(v) = p_v + \int d^3x\, v \cdot \nabla\psi_v(x)\, \nabla\psi_v(x), \quad p_v = v/(1-v^2)^{1/2}. \quad (2.1.12)$$

We will prove that (ψ_v, π_v) is the unique critical point and, moreover, the global minimum of $\mathcal{H}_{P(v)}$. Thus, if initial data are close to (ψ_v, π_v), then the corresponding solution must remain close forever by conservation of energy, which translates into the orbital stability of the solitons. Here we follow the ideas of the D. Bambusi and L. Galgani paper [99], where the orbital stability of solitons for the Maxwell–Lorentz equations was proved for the first time. For a general class of nonlinear wave equations with symmetries, such an approach to orbital stability of the solitons was developed in [100, 101].

However, the orbital stability by itself is not enough. It only ensures that initial states, close to a soliton, remain so; it does not yield the convergence of $\dot{q}(t)$ in (2.1.8), and even less the asymptotics (2.1.9), (2.1.10). Thus, we need an additional, not-quite-obvious argument that combines the relaxation (1.5.17) with the orbital stability to establish the soliton-like asymptotics (2.1.8), (2.1.9), (2.1.10). As one essential input, we will use the strong Huygens principle for wave equation.

2.1.1 Canonical Transformation and Reduced System

Since the total momentum is conserved, it is natural to use P as a new coordinate. To maintain the symplectic structure, we have to complete this coordinate to a canonical transformation of the Hilbert phase space $\mathcal{E}$.

Definition 2.1.2 Let the transform $T : \mathcal{E} \to \mathcal{E}$ be defined by

$$T : Y = (\psi, \pi, q, p) \mapsto Y^T = (\Psi(x), \Pi(x), Q, P)$$
$$= (\psi(q + x), \pi(q + x), q, P(\psi, q, \pi, p)), \qquad (2.1.13)$$

where $P(\psi, q, \pi, p)$ is the total momentum (2.1.2).

Remarks 2.1.3 (i) The map T is continuous on $\mathcal{E}$ and Fréchet-differentiable at points $Y = (\psi, q, \pi, p)$ with sufficiently smooth $\psi(x), \pi(x)$, but it is not everywhere differentiable.
(ii) In the T-coordinates the solitons $Y_{v,a}(t) = (\psi_v(x - a - vt), \pi_v(x - a - vt), q = a + vt, p_v)$ become stationary, except for the coordinate Q,

$$T Y_{v,a}(t) = (\psi_v(x), \pi_v(x), a + vt, P(v)), \qquad (2.1.14)$$

with the total momentum $P(v)$ of the soliton defined in (2.1.12).

Denote $\mathcal{H}^T(Y) = \mathcal{H}(T^{-1}Y)$ for $Y = (\Psi, \Pi, Q, P) \in \mathcal{E}$. Then

$$\mathcal{H}^T(\Psi, \Pi, Q, P) = \mathcal{H}_P(\Psi, \Pi)$$
$$= \mathcal{H}(\Psi(x - Q), \Pi(x - Q), Q, P + \int d^3x\, \Pi(x)\, \nabla\Psi(x))$$
$$= \int d^3x \left[\frac{1}{2}|\Pi(x)|^2 + \frac{1}{2}|\nabla\Psi(x)|^2 + \Psi(x)\rho(x) \right]$$
$$+ \left(1 + \left[P + \int d^3x\, \Pi(x)\, \nabla\Psi(x)\right]^2\right)^{1/2}.$$

The functionals $\mathcal{H}^T$ and $\mathcal{H}$ are Fréchet-differentiable on the Hilbert phase space $\mathcal{E}$.

Proposition 2.1.4 *Let $Y(t) \in C(\mathbb{R}, \mathcal{E})$ be a solution to the system (2.1.1). Then*

$$Y^T(t) := TY(t) = (\Psi(t), \Pi(t), P(t), Q(t)) \in C(\mathbb{R}, \mathcal{E})$$

is a solution to the Hamiltonian system

$$\begin{cases} \dot{\Psi} = D_\Pi \mathcal{H}^T, & \dot{\Pi} = -D_\Psi \mathcal{H}^T \\ \dot{Q} = D_P \mathcal{H}^T, & \dot{P} = -D_Q \mathcal{H}^T \end{cases}. \qquad (2.1.15)$$

Proof The equations for $\dot{\Psi}$, $\dot{\Pi}$, and $\dot{Q}$ can be checked by direct computation, while the one for $\dot{P}$ follows from conservation of the total momentum (2.1.2) since the Hamiltonian $\mathcal{H}^T$ does not depend on Q. $\qquad \square$

Remark 2.1.5 Formally, Proposition 2.1.4 follows from the fact that T is a canonical transform; see Section 2.1.4.

Recall that Q is a cyclic coordinate. Hence, the system (2.1.15) is equivalent to a reduced Hamiltonian system for Ψ and Π only, which can be written as

$$\dot{\Psi} = D_\Pi \mathcal{H}_P, \qquad \dot{\Pi} = -D_\Psi \mathcal{H}_P. \qquad (2.1.16)$$

Due to (2.1.14), the soliton (ψ_v, π_v) is a stationary solution to (2.1.16) with $P = P(v)$. Moreover, for every fixed $P \in \mathbb{R}^3$, the functional $\mathcal{H}_P$ is Fréchet-differentiable on the Hilbert phase space $\mathcal{F} = \overset{\circ}{H}{}^1 \oplus L^2$. Hence, (2.1.16) implies that the soliton is a critical point of $\mathcal{H}_{P(v)}$ on $\mathcal{F}$. The next lemma demonstrates that (ψ_v, π_v) is a global minimum of $\mathcal{H}_{P(v)}$ on $\mathcal{F}$.

Lemma 2.1.6 *(i) For every $v \in \mathbb{R}^3$ with $|v| < 1$, the functional $\mathcal{H}_{P(v)}$ has the lower bound*

$$\mathcal{H}_{P(v)}(\Psi, \Pi) - \mathcal{H}_{P(v)}(\psi_v, \pi_v)$$
$$\geq \frac{1 - |v|}{2} \left(\|\nabla(\Psi - \psi_v)\|^2 + \|\Pi - \pi_v\|^2 \right), \quad (\Psi, \Pi) \in \mathcal{F}. \qquad (2.1.17)$$

(ii) $\mathcal{H}_{P(v)}$ has no other critical points on $\mathcal{F}$, except the point (ψ_v, π_v).

Proof (i) Denoting $\Psi - \psi_v = \psi$ and $\Pi - \pi_v = \pi$, we have

$$\mathcal{H}_{P(v)}(\psi_v + \psi, \pi_v + \pi) - \mathcal{H}_{P(v)}(\psi_v, \pi_v)$$
$$= \int d^3 x (\pi_v(x)\pi(x) + \nabla\psi_v(x) \cdot \nabla\psi(x) + \rho(x)\psi(x))$$
$$+ \frac{1}{2} \int d^3 x \left(|\nabla\psi(x)|^2 + |\pi(x)|^2 \right) + (1 + (p_v + m)^2)^{1/2} - (1 + p_v^2)^{1/2},$$

$$(2.1.18)$$

where $p_v = P(v) + \int d^3x\, \pi_v(x)\, \nabla\psi_v(x)$, and

$$m = \int d^3x\, (\pi(x)\, \nabla\psi_v(x) + \pi_v(x)\, \nabla\psi(x) + \pi(x)\, \nabla\psi(x)).$$

Taking into account that $v = (1 + p_v^2)^{-1/2} p_v$, we obtain

$$\mathcal{H}_{P(v)}(\psi_v + \psi, \pi_v + \pi) - \mathcal{H}_{P(v)}(\psi_v, \pi_v)$$

$$= \frac{1}{2}\int d^3x\, (|\pi(x)|^2 + |\nabla\psi(x)|^2) + (1 + p_v^2)^{-1/2}\int d^3x\, \pi(x)\, p_v \cdot \nabla\psi(x)$$

$$- (1 + p_v^2)^{-1/2} p_v \cdot m + (1 + (p_v + m)^2)^{1/2} - (1 + p_v^2)^{1/2}.$$

It is easy to check that the expression in the third line is non-negative. Then the lower bound (2.1.17) follows by using $|(1 + p_v^2)^{-1/2} p_v| = |v|$.

(ii) If $(\Psi, \Pi) \in \mathcal{F}$ is a critical point for $\mathcal{H}_{P(v)}$, then it satisfies

$$0 = \Pi(x) + (1 + \tilde{p}^2)^{-1/2}\tilde{p} \cdot \nabla\Psi(x),$$

$$0 = -\Delta\Psi(x) + \rho(x) - (1 + \tilde{p}^2)^{-1/2}\tilde{p} \cdot \nabla\Pi(x),$$

where $\tilde{p} = P(v) + \int d^3x\, \Pi(x)\, \nabla\Phi(x)$. This system is equivalent to equation (2.1.5) for solitons in the case of the velocity $\tilde{v} = (1 + \tilde{p}^2)^{-1/2}\tilde{p}$. Hence, $\Psi = \psi_{\tilde{v}}$, $\Pi = \pi_{\tilde{v}}$, and $P(\tilde{v}) = P(v)$.

It remains to check that $\tilde{v} = v$. Indeed, for the total momentum $P(v)$ of the soliton solution (2.1.3), the Parseval identity and (2.1.6) imply that

$$P(v) = p_v + \int d^3x\, v \cdot \nabla\psi_v(x)\, \nabla\psi_v(x)$$

$$= \frac{v}{\sqrt{1 - v^2}} + (2\pi)^{-3}\int d^3k\, \frac{(v \cdot k)\hat{\rho}(k)\overline{k\hat{\rho}(k)}}{(k^2 - (v \cdot k)^2)^2}.$$

Hence, $P(v) = \varkappa(|v|)v$ with $\varkappa(|v|) \ge 0$, and for $v \ne 0$, one has

$$|P(v)| = \frac{|v|}{\sqrt{1 - v^2}} + \frac{1}{(2\pi)^3|v|}\int d^3k\, \frac{|(v \cdot k)\hat{\rho}(k)|^2}{(k^2 - (v \cdot k)^2)^2}.$$

Since $|P(v)| = \varkappa(|v|)|v|$ is a monotone increasing function of $|v| \in [0, 1)$, we conclude that $v = \tilde{v}$. $\qquad\square$

Remark 2.1.7 Proposition 2.1.4 is not really needed for the proof of Theorem 2.1.1. However, the proposition together with (2.1.14) and (2.1.16) shows that (ψ_v, π_v) is a critical point and suggests an investigation of the stability through a lower bound, as in (2.1.17). In Section 2.1.4 we sketch the derivation of Proposition 2.1.4 for sufficiently smooth solutions based only on the invariance of symplectic structure. We expect that a similar proposition holds for other translation-invariant systems similar to (2.1.1).

2.1.2 Orbital Stability of Solitons

We follow [99] deducing orbital stability from the conservation of the Hamiltonian $\mathcal{H}_P$ together with its lower bound (2.1.17). For $|v| < 1$, denote

$$\delta = \delta(v) = \|\psi^0(x) - \psi_v(x - q^0)\| + \|\pi^0(x) - \pi_v(x - q^0)\| + |p^0 - p_v|. \tag{2.1.19}$$

Lemma 2.1.8 *Let $Y(t) = (\psi(t), \pi(t), q(t), p(t)) \in C(\mathbb{R}, \mathcal{E})$ be a solution to (2.1.1) with an initial state $Y(0) = Y^0 = (\psi^0, \pi^0, q^0, p^0) \in \mathcal{E}$. Then, for every $\varepsilon > 0$, there exists a $\delta_\varepsilon > 0$ such that*

$$\|\psi(q(t) + x, t) - \psi_v(x)\|$$
$$+ \|\pi(q(t) + x, t) - \pi_v(x)\| + |p(t) - p_v| \leq \varepsilon, \qquad t \in \mathbb{R}, \tag{2.1.20}$$

provided $\delta \leq \delta_\varepsilon$.

Proof Denote by P^0 the total momentum of the considered solution $Y(t)$. There exists a soliton solution (2.1.3) corresponding to some velocity $\tilde{v}$ with the same total momentum $P(\tilde{v}) = P^0$. Then, (2.1.19) implies that $|P^0 - P(v)| = |P(\tilde{v}) - P(v)| = \mathcal{O}(\delta)$. Hence, also, $|\tilde{v} - v| = \mathcal{O}(\delta)$ and

$$\|\psi^0(x) - \psi_{\tilde{v}}(x - q^0)\| + \|\pi^0(x) - \pi_{\tilde{v}}(x - q^0)\| + |p^0 - p_{\tilde{v}}| = \mathcal{O}(\delta).$$

Therefore, denoting $(\Psi^0, \Pi^0, Q^0, P^0) = TY^0$, we have

$$\mathcal{H}_{P(\tilde{v})}(\Psi^0, \Pi^0) - \mathcal{H}_{P(\tilde{v})}(\psi_{\tilde{v}}, p_{\tilde{v}}) = \mathcal{O}(\delta^2). \tag{2.1.21}$$

Total momentum and energy conservation imply that for $(\Psi(t), \Pi(t),$ $Q(t), P^0) = TY(t)$,

$$\mathcal{H}_{P(\tilde{v})}(\Psi(t), \Pi(t)) = \mathcal{H}(TY(t)) = \mathcal{H}_{P(\tilde{v})}(\Psi^0, \Pi^0) \text{ for } t \in \mathbb{R}.$$

Hence, (2.1.21) and (2.1.17) with $\tilde{v}$ instead of v imply that

$$\|\Psi(t) - \psi_{\tilde{v}}\| + \|\Pi(t) - \pi_{\tilde{v}}\| = \mathcal{O}(\delta) \tag{2.1.22}$$

uniformly in $t \in \mathbb{R}$. On the other hand, total momentum conservation implies that

$$p(t) = P(\tilde{v}) + \langle \Pi(t), \nabla\Psi(t) \rangle \text{ for } t \in \mathbb{R}.$$

Therefore, (2.1.22) leads to

$$|p(t) - p_{\tilde{v}}| = \mathcal{O}(\delta) \tag{2.1.23}$$

uniformly in $t \in \mathbb{R}$. Finally, (2.1.22), (2.1.23) together imply (2.1.20) because $|\tilde{v} - v| = \mathcal{O}(\delta)$. $\qquad\square$

2.1.3 Strong Huygens Principle and Soliton Asymptotics

We combine the relaxation of the acceleration and orbital stability with the strong Huygens principle to prove Theorem 2.1.1.

Proposition 2.1.9 *Let the assumptions of Theorem 2.1.1 be fulfilled. Then, for every $\delta > 0$, there exist a $t_* = t_*(\delta)$ and a solution $Y_*(t) = (\psi_*(x,t), \pi_*(x,t), q_*(t), p_*(t)) \in C([t_*, \infty), \mathcal{E})$ to the system (2.1.1) such that*

(i) $Y_*(t)$ coincides with $Y(t)$ in the future cone,

$$q_*(t) = q(t) \qquad \qquad \text{for } t \geq t_*, \qquad \qquad (2.1.24)$$

$$\psi_*(x,t) = \psi(x,t) \qquad \text{for } |x - q(t_*)| < t - t_*. \qquad (2.1.25)$$

(ii) $Y_*(t_*)$ is close to a soliton $Y_{v,a}$ with some v and a,

$$\|Y_*(t_*) - Y_{v,a}\|_{\mathcal{E}} \leq \delta. \qquad \qquad (2.1.26)$$

Proof The Kirchhoff formula gives

$$\psi(x,t) = \psi_r(x,t) + \psi_0(x,t), \qquad x \in \mathbb{R}^3, \ t > 0,$$

where

$$\psi_r(x,t) = -\int \frac{d^3 y}{4\pi |x - y|} \rho(y - q(t - |x - y|)), \qquad (2.1.27)$$

$$\psi_0(x,t) = \frac{1}{4\pi t} \int_{S_t(x)} d^2 y \, \pi(y,0) + \frac{\partial}{\partial t} \left(\frac{1}{4\pi t} \int_{S_t(x)} d^2 y \, \psi(y,0) \right). \qquad (2.1.28)$$

Here $S_t(x)$ denotes the sphere $|y - x| = t$. Let us assume for simplicity that initial fields vanish. The general case can easily be reduced to this situation using the strong Huygens principle. We will comment on this reduction at the end of the proof.

In the case of zero initial data the solution reduces to the retarded potential:

$$\psi(x,t) = \psi_r(x,t), \quad x \in \mathbb{R}^3, \quad t > 0.$$

We construct the solution $Y_*(t)$ as a modification of $Y(t)$. First, we modify the trajectory $q(t)$. The relaxation of acceleration (2.1.8) means that for any $\varepsilon > 0$, there exist $t_\varepsilon > 0$ such that

$$|\ddot{q}(t)| \leq \varepsilon, \qquad t \geq t_\varepsilon.$$

Hence, the trajectory for large times locally tends to a straight line; i.e., for any fixed $T > 0$,

$$q(t) = q(t_\varepsilon) + (t - t_\varepsilon)\dot{q}(t_\varepsilon) + r(t_\varepsilon, t), \quad \text{where}$$

$$\max_{t \in [t_\varepsilon, t_\varepsilon + T]} |r(t_\varepsilon, t)| \to 0, \quad t_\varepsilon \to \infty.$$

Denote $\lambda_c(t) := q(t_\varepsilon) + \dot{q}(t_\varepsilon)(t - t_\varepsilon)$ and define the modified trajectory as

$$q_*(t) = \begin{cases} \lambda_\varepsilon(t), & t \le t_\varepsilon \\ q(t), & t \ge t_\varepsilon \end{cases}. \tag{2.1.29}$$

Then

$$\ddot{q}_*(t) = \begin{cases} 0, & t < t_\varepsilon \\ \ddot{q}(t), & t > t_\varepsilon \end{cases}.$$

In the next step, we define the modified field as retarded potential of type (2.1.27):

$$\psi_*(x,t) = -\int \frac{d^3 y}{4\pi |x - y|} \rho(y - q_*(t - |x - y|)), \quad x \in \mathbb{R}^3, \quad t \in \mathbb{R}. \tag{2.1.30}$$

Lemma 2.1.10 *The RHS of (2.1.30) depends on the trajectory $q_*(\tau)$ only from a bounded interval of time $\tau \in [t - T(x,t), t]$, where*

$$T(x,t) := \frac{R_\rho + |x - q(t)|}{1 - \overline{v}}. \tag{2.1.31}$$

Here $\overline{v} = \sup\limits_{t \in \mathbb{R}} |\dot{q}(t)| < 1$ by (1.5.11).

Proof This lemma is obvious geometrically, and its formal proof also is easy. The integrand of (2.1.30) vanishes for $|y - q_*(t - |x - y|)| \ge R_\rho$ by (1.5.14). Therefore, the integral is spread over the region $|y - q_*(t - |x - y|)| \le R_\rho$, which implies $|y - q_*(t) + q_*(t) - q_*(t - |x - y|)| \le R_\rho$. Hence,

$$|y - q_*(t)| \le R_\rho + \overline{v}|x - y|.$$

On the other hand, $|x - y| \le |x - q_*(t)| + |y - q_*(t)|$, and hence,

$$|y - q_*(t)| \ge -|x - q_*(t)| + |x - y|.$$

Therefore,

$$-|x - q_*(t)| + |x - y| \le R_\rho + \overline{v}|x - y|,$$

which implies

$$|x - y| \le \frac{R_\rho + |x - q_*(t)|}{1 - \overline{v}}.$$

Now the lemma is proved. $\qquad\square$

The potential (2.1.30) satisfies the wave equation

$$\ddot{\psi}_*(x,t) = \Delta\psi_*(x,t) - \rho(x - q_*(t)), \quad x \in \mathbb{R}^3, \quad t \in \mathbb{R}.$$

We should still prove equations for the trajectory $q_*(t)$:

$$\dot{q}_*(t) = \frac{p_*(t)}{\sqrt{1 + p_*^2(t)}}, \qquad \dot{p}_*(t) = -\int \nabla\psi_*(x,t)\rho(x - q_*(t))\,dx, \qquad t > t_*$$

$$\tag{2.1.32}$$

with sufficiently large $t_* \geq t_\varepsilon$. Let us note that the integral here is spread over the ball $|x - q_*(t)| \leq R_\rho$. Now Lemma 2.1.10 implies that $\psi_*(x,t)$ depends on the trajectory $q_*(\tau)$ only from a bounded interval $\tau \in [t - \overline{T}, t]$, where

$$\overline{T} := \frac{2R_\rho}{1 - \overline{v}}.$$

Let us define $t_* := t_\varepsilon + \overline{T}$. Then, by Lemma 2.1.10,

$$\psi_*(x,t) = \psi(x,t), \quad t > t_*, \quad |x - q_*(t)| \leq R_\rho$$

since $q_*(t) \equiv q(t)$ for $t > t_* - \overline{T} = t_\varepsilon$ by (2.1.29). Hence, equations (2.1.32) hold for $q_*(t)$ as well as for $q(t)$.

It remains to prove (2.1.26). The key observation is that outside the cone $K_\varepsilon := \{(x,t) \in \mathbb{R}^4 : |x - q(t_\varepsilon)| < t - t_\varepsilon\}$, the retarded potential (2.1.30) coincides with the soliton $\psi_{v,a}(x,t)$, where $v = \dot{q}(t_\varepsilon)$ and $a = q(t_\varepsilon)$ by our definition (2.1.29). In particular,

$$\psi(x,t_*) = \psi_{v,a}(x - a - vt_*), \qquad |x - q(t_\varepsilon)| > t_* - t_\varepsilon = \overline{T}.$$

In the ball $|x - q(t_*)| < \overline{T}$, the coincidence generally does not hold, but the difference of the LHS with the RHS converges to zero as $\varepsilon \to 0$ uniformly for $|x - q(t_*)| < \overline{T}$, and such uniform convergence holds for the gradient of the difference. This follows from the integral representation (2.1.30) by Lemma 2.1.10, since

$$\max_{t \in (t_* - T(x,t_*), t_*)} [|q_*(t) - \lambda_\varepsilon(t)| + |\dot{q}_*(t) - \dot{\lambda}_\varepsilon(t)|] \to 0, \qquad \varepsilon \to 0$$

by the relaxation of acceleration (2.1.8). It is important that $T(x,t_*)$ is bounded for $|x - q(t_*)| < \overline{T}$ by (2.1.31). This proves Proposition 2.1.9 in the case of zero initial data.

The next step is the proof for initial data with bounded support:

$$\psi(x,0) = \pi(x,0) = 0, \qquad |x| > R_0.$$

Now we apply the strong Huygens principle: in this case the potential (2.1.28) vanishes in a future cone,

$$\psi_0(x,t) = 0, \qquad |x| < t - R_0.$$

However, the estimate $|\dot{q}(t)| \leq \bar{v} < 1$ implies that the trajectory $(q(t),t)$ lies in this cone for all $t > t_0$. Hence, the solution for $t > t_0$ again reduces to the retarded potential, and the needed conclusion follows.

Finally, arbitrary finite-energy initial data admits a splitting in two summands: the first vanishing for $|x| > R_0$ and the second vanishing for $|x| < R_0 - 1$. The energy of the second summand is arbitrarily small for large R_0, and the energy of the corresponding potential (2.1.28) is conserved in time since it is a solution to the free wave equation. Hence, its role is negligible for sufficiently large R_0. □

Now we can prove our main result.

Proof of Theorem 2.1.1 For every $\varepsilon > 0$, there exists $\delta > 0$ such that (2.1.26) implies by Lemma 2.1.8 that

$$\|\psi_*(q_*(t) + x,t) - \psi_v(x)\| + \|\pi_*(q_*(t) + x,t) - \pi_v(x)\|$$
$$+ |\dot{q}_*(t) - v| \leq \varepsilon \text{ for } t > t_*.$$

Therefore, (2.1.24) and (2.1.25) imply that for every $R > 0$ and $t > t_* + \frac{R}{1-\bar{v}}$,

$$\|\psi(q(t) + x, t) - \psi_v(x)\|_R + \|\pi(q(t) + x, t) - \pi_v(x)\|_R + |\dot{q}(t) - v|$$
$$= \|\psi_*(q_*(t) + x,t) - \psi_v(x)\|_R + \|\pi_*(q_*(t) + x,t) - \pi_v(x)\|_R$$
$$+ |\dot{q}_*(t) - v| \leq \varepsilon.$$

Since $\varepsilon > 0$ is arbitrary, we conclude (2.1.10). Theorem 2.1.1 is proved.

2.1.4 Invariance of Symplectic Structure

The canonical equivalence of the Hamiltonian systems (2.1.1) and (2.1.15) can be seen from the Lagrangian viewpoint. We remain at the formal level. For a complete mathematical justification we would have to develop some theory of infinite-dimensional Hamiltonian systems, which is beyond the scope of this book.

By definition we have $\mathcal{H}^T(\Psi, \Pi, Q, P) = \mathcal{H}(\psi, \pi, q, p)$, with the arguments related through the transformation T. To each Hamiltonian, we associate a Lagrangian through the Legendre transformation

$$L(\psi, \dot{\psi}, q, \dot{q}) = \langle \pi, \dot{\psi} \rangle + p \cdot \dot{q} - \mathcal{H}(\psi, \pi, q, p) , \qquad \dot{\psi} = D_\pi \mathcal{H} , \quad \dot{q} = D_p \mathcal{H} ,$$

$$L^T(\Psi, \dot{\Phi}, Q, \dot{Q}) = \langle \Pi, \dot{\Psi} \rangle + P \cdot \dot{Q} - \mathcal{H}^T(\Psi, \Pi, Q, P) , \quad \dot{\Psi} = D_\Pi \mathcal{H}^T, \quad \dot{Q} = D_P \mathcal{H}^T .$$

These Legendre transforms are well defined because the Hamiltonian functionals are convex in the momenta.

Lemma 2.1.11 *The following identity holds:*

$$L^T(\Psi, \dot{\Psi}, Q, \dot{Q}) = L(\psi, \dot{\psi}, q, \dot{q}).$$

Proof Clearly we have to check the invariance of the canonical form,

$$\langle \Pi, \dot{\Psi} \rangle + P \cdot \dot{Q} = \langle \pi, \dot{\psi} \rangle + p \cdot \dot{q}. \tag{2.1.33}$$

For this purpose we substitute

$$\begin{cases} \Pi(x) = \pi(q + x), & \dot{\Psi}(x) = \dot{\psi}(q + x) + \dot{q} \cdot \nabla\psi(q + x) \\ P = p - \displaystyle\int \dot{\psi} \cdot \nabla\psi \, dx, & \dot{Q} = \dot{q} \end{cases} .$$

Then the LHS of (2.1.33) becomes

$$\langle \pi(q + x), \dot{\psi}(q + x) + \dot{q} \cdot \nabla\psi(q + x) \rangle$$

$$+ (p - \langle \pi(x), \nabla\psi(x) \rangle) \cdot \dot{q} = \langle \pi, \dot{\psi} \rangle + p \cdot \dot{q}.$$

The lemma is proved. ▫

This lemma implies that the corresponding action functionals are identical when transformed by T. Hence, finally, the two Hamiltonian systems (2.1.1) and (2.1.15) are equivalent since dynamical trajectories are stationary points of the respective action functionals.

2.1.5 Translation-Invariant Maxwell–Lorentz System

In [55], asymptotics of type (2.1.8)–(2.1.10) were extended to the Maxwell–Lorentz translation-invariant system (1.6.1) without external fields. In this case, the Hamiltonian coincides with (1.6.3), where $V(x) \equiv 0$. The extension of methods [54] to this case required a new detailed analysis of the corresponding Hamiltonian structure which is necessary for the canonical transformation. Now the key role in applying the strong Huygens principle is played by new estimates of long-time decay for oscillations of energy and total momentum for solutions of the perturbed Maxwell–Lorentz system (estimates (4.24)–(4.25) in [55]).

2.2 The Case of Weak Coupling

In [57] the soliton asymptotic of type (2.1.8)–(2.1.10) for the system (1.5.1)–(1.5.2) was proved in a stronger form for the case of a weak coupling

$$\|\rho\|_{L^2(\mathbb{R}^3)} \ll 1. \tag{2.2.1}$$

Namely, in [57], initial fields are considered with decay $|x|^{-5/2-\varepsilon}$, where $\varepsilon > 0$ (condition (2.2) in [57]), provided that $\nabla V(q) = 0$ for $|q| > $ const. Under these assumptions, more strong decay holds,

$$|\ddot{q}(t)| \le C(1 + |t|)^{-1-\varepsilon}, \qquad t \in \mathbb{R} \tag{2.2.2}$$

for "outgoing solutions" that satisfy the condition

$$|q(t)| \to \infty, \qquad t \to \pm\infty. \tag{2.2.3}$$

With these assumptions, asymptotics (2.1.8)–(2.1.10) can be significantly strengthened: now

$$\dot{q}(t) \to v_{\pm}, \qquad (\psi(x,t), \pi(x,t)) = (\psi_{v_{\pm}}(x - q(t)), \pi_{v_{\pm}}(x - q(t)))$$
$$+ W(t)\Phi_{\pm} + (r_{\pm}(x,t), s_{\pm}(x,t)),$$

where "dispersive waves" $W(t)\Phi_{\pm}$ are solutions of a free wave equation shown in Figure 2.1.

Now the remainder converges to zero in *global energy norm*:

$$\|\nabla r_{\pm}(q(t),t)\| + \|r_{\pm}(q(t),t)\| + \|s_{\pm}(q(t),t)\| \to 0, \qquad t \to \pm\infty.$$

This progress compared with local decay (2.1.10) is due to the fact that we identified a dispersive wave $W(t)\Phi_{\pm}$ under the condition of smallness (2.2.1). This identification is possible due to the rapid decay (2.2.2), in difference with (1.5.17).

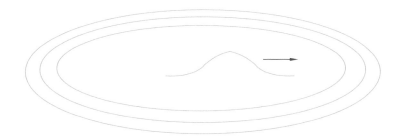

Figure 2.1 Soliton and dispersive waves.

All solitons propagate with velocities $v < 1$, and therefore they are spatially separated for a large time from the dispersive waves $W(t)\Phi_\pm$, which propagate with unit velocity (Figure 2.1).

The proofs rely on integral Duhamel representation and on rapid dispersive decay of solutions to the free wave equation. Similar results were obtained in [126] for a system of type (1.5.1)–(1.5.2) with the Klein–Gordon equation and in [56] for the Maxwell–Lorentz equations (1.6.1) with the same smallness condition (2.2.3) under the assumption that $E^{\text{ext}}(x) = B^{\text{ext}}(x) = 0$ for $|x| >$ const. In [88], this result was extended to the Maxwell–Lorentz equations of type (1.6.1) with a rotating charge.

Remark 2.2.1 The results of [57, 88] imply A. Soffer's "Grand Conjecture" [161, p. 460] in a moving frame for translation-invariant systems under the condition of smallness (2.2.1).

Open problem Global attraction to solitons for the *relativistic* nonlinear wave equations

$$\ddot{\psi}(x,t) = -\Delta \psi(x,t) + f(\psi(x,t)), \qquad x \in \mathbb{R}^n \qquad (2.2.4)$$

is still an open problem. Numerical simulations [58] for the case $n = 1$ confirm the asymptotics (13) for a broad class of the nonlinearities (see Chapter 6).

3

Global Attraction to Stationary Orbits

The global attraction to stationary orbits (15) was proved (i) in [61]–[67] for the Klein–Gordon and Dirac equations coupled to $U(1)$-invariant nonlinear oscillators; (ii) in [60] for discrete in space and time difference approximations of such coupled systems, i.e., for the corresponding difference schemes; and (iii) in [50] and [69]–[71] for the wave, Klein–Gordon and Dirac equations with concentrated nonlinearities.

The attraction (15) was proved under the assumption that the equations are *strictly nonlinear*. For linear equations, the attraction can fail if the discrete spectrum consists of at least two points.

In this chapter, we present with detail the first results on global attraction to stationary orbits (15) obtained in [62]–[64]. The results concern the global attraction for a 1D Klein–Gordon equation coupled to a nonlinear oscillator.

The proofs of these results rely on (i) the concept of omega-limit trajectory, (ii) a nonlinear analog of the Kato theorem on the absence of embedded eigenvalues, (iii) new theory of multipliers in the space of quasimeasures, and (iv) novel application of the Titchmarsh convolution theorem.

Besides the formal proof, in Section 3.10 we give an informal explanation of the *nonlinear radiative mechanism*, which causes the global attraction.

In conclusion, we specify the general conjecture on global attractors (6), which summarizes all results of Chapters 1, 2, and 3 (see Section 3.11).

3.1 Nonlinear Klein–Gordon Equation

The first results on global attraction to stationary orbits (15) were established in [62]–[64] for the Klein–Gordon equation coupled to a nonlinear oscillator:

$$\ddot{\psi}(x,t) = \psi''(x,t) - m^2\psi(x,t) + \delta(x)F(\psi(0,t)), \qquad x \in \mathbb{R}, \quad t \in \mathbb{R}.$$

$$(3.1.1)$$

The asymptotics (6) for this equation mean that

$$\psi(x,t) \sim \psi_{\pm}(x)e^{-i\omega_{\pm}t}, \qquad t \to \pm\infty. \tag{3.1.2}$$

We consider complex solutions, identifying complex values $\psi \in \mathbb{C}$ with the vectors $(\psi_1, \psi_2) \in \mathbb{R}^2$, where $\psi_1 = \operatorname{Re}\psi$ and $\psi_2 = \operatorname{Im}\psi$. Suppose that $F \in C^1(\mathbb{R}^2, \mathbb{R}^2)$ and

$$F(\psi) = -\nabla_{\overline{\psi}}U(\psi), \qquad \psi \in \mathbb{C}, \tag{3.1.3}$$

where U is a real function and $\nabla_{\overline{\psi}} := (\partial_1, \partial_2)$. In this case, equation (3.1.5) is formally equivalent to the Hamiltonian system of type (1.1.2) in the Hilbert phase space $\mathcal{E} := H^1(\mathbb{R}) \oplus L^2(\mathbb{R})$. The Hamiltonian functional is

$$\mathcal{H}(\psi, \pi) = \frac{1}{2} \int \left[|\pi(x)|^2 + |\psi'(x)|^2 + m^2|\psi(x)|^2 \right] dx$$
$$+ U(\psi(0)), \qquad (\psi, \pi) \in \mathcal{E}. \tag{3.1.4}$$

Let us write (3.1.1) in the vector form as

$$\dot{Y}(t) = \mathcal{F}(Y(t)), \qquad t \in \mathbb{R}, \tag{3.1.5}$$

where $Y(t) = (\psi(t), \dot{\psi}(t))$. We assume that

$$\inf_{\psi \in \mathbb{C}} U(\psi) > -\infty. \tag{3.1.6}$$

In this case, a finite-energy solution $Y(t) \in C(\mathbb{R}, \mathcal{E})$ exists and is unique for any initial state $Y(0) \in \mathcal{E}$ (see Appendix C of [64] for details). The a priori bound

$$\sup_{t \in \mathbb{R}}[\|\dot{\psi}(t)\|_{L^2(\mathbb{R})} + \|\psi(t)\|_{H^1(\mathbb{R})}] < \infty \tag{3.1.7}$$

holds due to the conservation of energy (3.1.4). Note that the confining condition of type (1.5.7) is no longer necessary, since conservation of energy (3.1.4) with $m > 0$ ensures the boundedness of solutions.

Furthermore, we assume the $U(1)$-invariance of the potential:

$$U(\psi) = u(|\psi|), \qquad \psi \in \mathbb{C}. \tag{3.1.8}$$

Then the differentiation in (3.1.3) gives us that

$$F(\psi) = a(|\psi|)\psi, \qquad \psi \in \mathbb{C}, \tag{3.1.9}$$

and therefore

$$F(e^{i\theta}\psi) = e^{i\theta}F(\psi), \qquad \theta \in \mathbb{R}. \tag{3.1.10}$$

By "stationary orbits," we mean solutions of the form

$$\psi(x,t) = \psi_\omega(x)e^{-i\omega t} \tag{3.1.11}$$

with $\omega \in \mathbb{R}$ and $\psi_\omega \in H^1(\mathbb{R})$. Each stationary orbit corresponds to some solution of the equation

$$-\omega^2 \psi_\omega(x) = \psi_\omega''(x) - m^2 \psi_\omega(x) + \delta(x)F(\psi_\omega(0)), \qquad x \in \mathbb{R}, \tag{3.1.12}$$

which is the *nonlinear eigenvalue problem*. Any solution $\psi_\omega \in H^1(\mathbb{R})$ of this equation has the form

$$\psi_\omega(x) = Ce^{-\kappa|x|}, \qquad \kappa := \sqrt{m^2 - \omega^2} > 0, \tag{3.1.13}$$

and the constant C satisfies the nonlinear algebraic equation

$$2\kappa C = F(C). \tag{3.1.14}$$

Hence, the solutions $\psi_\omega \in H^1(\mathbb{R})$ exist for ω in some subset $\Omega \subset \mathbb{R}$ lying in the *spectral gap* $[-m, m]$. We denote the corresponding *solitary manifold* by $\mathcal{S}$:

$$\mathcal{S} = \{(e^{i\theta}\psi_\omega, -i\omega e^{i\theta}\psi_\omega) \in \mathcal{E} : \omega \in \Omega, \ \theta \in [0, 2\pi]\}. \tag{3.1.15}$$

Finally, suppose that the equation (3.1.5) is *strictly nonlinear*:

$$U(\psi) = u(|\psi|^2) = \sum_0^N u_j |\psi|^{2j}, \quad u_N > 0, \quad N \geq 2. \tag{3.1.16}$$

For example, the well-known *Ginzburg–Landau potential* $U(\psi) = |\psi|^4/4 - |\psi|^2/2$ satisfies all three conditions (3.1.6), (3.1.8), and (3.1.16).

Definition 3.1.1 (i) $\mathcal{E}_F \subset H^1_{\mathrm{loc}}(\mathbb{R}^3) \oplus L^2_{\mathrm{loc}}(\mathbb{R}^3)$ is the space $\mathcal{E}$ endowed with the seminorms

$$\|Y\|_{\mathcal{E},R} := \|Y\|_{H^1(-R,R)} + \|Y\|_{L^2(-R,R)}, \qquad R = 1, 2, \ldots. \tag{3.1.17}$$

(ii) Convergence in $\mathcal{E}_F$ is equivalent to convergence in all seminorms (3.1.17).

It is important to note that convergence in $\mathcal{E}_F$ is equivalent to convergence in a metric of type (1.2.9):

$$\mathrm{dist}[Y_1, Y_2] = \sum_{R=1}^\infty 2^{-R} \frac{\|Y_1 - Y_2\|_{\mathcal{E},R}}{1 + \|Y_1 - Y_2\|_{\mathcal{E},R}}, \qquad Y_1, Y_2 \in \mathcal{E}. \tag{3.1.18}$$

The main result of [62]–[64] is the following theorem.

Theorem 3.1.2 *Let the conditions (3.1.3), (3.1.6), (3.1.8), and (3.1.16) hold. Then any finite-energy solution* $Y(t) = (\psi(t), \dot\psi(t)) \in C(\mathbb{R}, \mathcal{E})$ *of (3.1.5) is attracted to the solitary manifold* (see Figure 2)*:*

$$Y(t) \xrightarrow{\mathcal{E}_F} \mathcal{S}, \qquad t \to \pm\infty, \qquad (3.1.19)$$

where the attraction is in the sense of (1.2.18).

3.2 Generalizations and Open Questions

Generalizations: The global attraction (3.1.19) was extended in [65] to the 1D Klein–Gordon equation with N nonlinear oscillators

$$\ddot\psi(x,t) = \psi''(x,t) - m^2\psi + \sum_{k=1}^{N} \delta(x - x_k) F_k(\psi(x_k,t)), \ x \in \mathbb{R}, \quad (3.2.1)$$

and in [61, 66, 67] it was extended to the Klein–Gordon and Dirac equations in $\mathbb{R}^n$ with $n \geq 3$ with a nonlocal interaction

$$\ddot\psi(x,t) = \Delta\psi(x,t) - m^2\psi + \sum_{k=1}^{N} \rho(x - x_k) F_k(\langle \psi(\cdot,t), \rho(\cdot - x_k)\rangle),$$

$$(3.2.2)$$

$$i\dot\psi(x,t) = (-i\alpha \cdot \nabla + \beta m)\psi + \rho(x) F(\langle \psi(\cdot,t), \rho\rangle), \qquad (3.2.3)$$

under the Wiener condition (1.5.13). Here $\alpha = (\alpha_1, \ldots, \alpha_n)$ and $\beta = \alpha_0$ are Dirac matrices.

Recently, the attraction (3.1.19) was extended in [69, 50, 70] to the 3D wave and Klein–Gordon equations with concentrated nonlinearities, and in [71] it was extended to the 1D Dirac equation coupled to a nonlinear oscillator.

In addition, the attraction (3.1.19) was extended in [60] to nonlinear space-time discrete Hamiltonian equations that are discrete approximations of equations of type (3.2.2), that is, that are the corresponding difference schemes. The proof relies on a new version of the Titchmarsh convolution theorem for distributions on a circle [68].

Open Questions:

I. Global attraction (3.1.2) to orbits with fixed frequencies $\omega_\pm$ has not yet been established.

II. Global attraction to stationary orbits for nonlinear Schrödinger equations has also not been established. In particular, such attraction is not proved for the 1D Schrödinger equation coupled to a nonlinear oscillator

$$i \dot{\psi}(x,t) = -\psi''(x,t) + \delta(x)F(\psi(0,t)), \qquad x \in \mathbb{R}. \tag{3.2.4}$$

The main difficulty is the infinite "spectral gap" $(-\infty, 0)$ (see Remark 3.9.3).

III. Global attraction to solitons (12) for the *relativistic* nonlinear Klein–Gordon equations is an open problem. In particular, it is open for the 1D equations (2.2.4).

3.3 Omega-Limit Trajectories

The proof of Theorem 3.1.2 is based on the general strategy of *omega-limit trajectories*, first introduced in [62] and developed further in [63]–[71].

Definition 3.3.1 An omega-limit trajectory for a given $Y(t) \in C(\mathbb{R}, \mathcal{E})$ is any limit function $Z(t)$ such that

$$Y(t + s_j) \xrightarrow{\mathcal{E}_F} Z(t), \qquad t \in \mathbb{R}, \tag{3.3.1}$$

as $s_j \to \infty$.

Definition 3.3.2 A function $Y(t) \in C(\mathbb{R}, \mathcal{E})$ is *omega-compact* if, for any sequence $s_j \to \infty$, there exists a subsequence $s_{j'} \to \infty$ such that (3.3.1) holds.

These concepts are useful in view of the following lemma, which lies at the basis of our approach.

Lemma 3.3.3 *Suppose that any solution $Y(t) \in C(\mathbb{R}, \mathcal{E})$ of (3.1.5) is omega-compact and any omega-limit trajectory is a stationary orbit,*

$$Z(x,t) = (\psi_\omega(x)e^{-i\omega t}, -i\omega\psi_\omega(x)e^{-i\omega t}), \tag{3.3.2}$$

where $\omega \in \mathbb{R}$. Then the attraction (3.1.19) holds for each solution $Y(t) \in C(\mathbb{R}, \mathcal{E})$ of (3.1.5).

Proof We need to show that

$$\lim_{t \to \infty} \operatorname{dist}(Y(t), \mathcal{S}) = 0.$$

Assume by contradiction that there exists a sequence $s_j \to \infty$ such that

$$\operatorname{dist}(Y(s_j), \mathcal{S}) \geq \delta > 0 \quad \forall j \in \mathbb{N}. \tag{3.3.3}$$

According to the omega-compactness of the solution Y, the convergence (3.3.1) holds for some subsequence $s_{j'} \to \infty$ and some stationary orbit (3.3.2):

$$Y(t + s_j) \xrightarrow{\mathcal{E}_F} Z(t), \qquad t \in \mathbb{R}. \tag{3.3.4}$$

But this convergence with $t = 0$ contradicts (3.3.3), since $Z(0) \in \mathcal{S}$ by definition (3.1.15). $\qquad\square$

For the proof of Theorem 3.1.2, it now suffices to check the assumptions of Lemma 3.3.3:

I. Each solution $Y(t) \in C(\mathbb{R}, \mathcal{E})$ of (3.1.5) is omega-compact.

II. Any omega-limit trajectory is a stationary orbit (3.3.2).

We check these conditions by analyzing the Fourier transform of solutions with respect to time. The main steps of the proof are as follows:

(1) Spectral representation for solutions to the nonlinear equation (3.1.5):

$$\psi(t) = \frac{1}{2\pi} \int e^{-i\omega t} \tilde{\psi}(\omega) d\omega. \qquad (3.3.5)$$

By the *spectrum* of a solution $\psi(t) := \psi(\cdot, t)$ we mean the support of its spectral density $\tilde{\psi}(\cdot)$, which is a tempered distribution of $\omega \in \mathbb{R}$ with values in H^1.

(2) The *absolute continuity* of the spectral density $\tilde{\psi}(\omega)$ *on the continuous spectrum* $(-\infty, -m) \cup (m, \infty)$ of the free Klein–Gordon equation. This is a nonlinear analog of the Kato theorem on the absence of embedded eigenvalues.

(3) The *omega-compactness* of each solution.

(4) The reduction of the spectrum *of each omega-limit trajectory* to a subset of the *spectral gap* $[-m, m]$.

(5) Reduction of this spectrum to a *single point* using the *Titchmarsh convolution theorem*.

Below we follow this program, referring at some points to the papers [62] and [64] for technically important properties of quasimeasures.

3.4 Limiting Absorption Principle

It suffices to prove the attraction (3.1.19) only for positive times. We split the solution into two summands

$$\psi(x, t) = \psi_0(x, t) + \psi_1(x, t), \qquad (3.4.1)$$

where

$$\begin{cases} \ddot{\psi}_0(x, t) = \psi_0''(x, t) - m^2 \psi_0(x, t), & \psi_0(x, 0) = \psi(x, 0), \ \dot{\psi}_0(x, 0) = \dot{\psi}(x, 0) \\ \ddot{\psi}_1(x, t) = \psi_1''(x, t) - m^2 \psi_1(x, t) + \delta(x) F(\psi(0, t)), & \psi_1(x, 0) = 0, \ \dot{\psi}_1(x, 0) = 0 \end{cases}$$

$$(x, t) \in \mathbb{R}^2. \qquad (3.4.2)$$

Both functions satisfy a priori bounds of type (3.1.7):

$$\sup_{t \in \mathbb{R}} [\|\dot{\psi}_k(t)\|_{L^2(\mathbb{R})} + \|\psi_k(t)\|_{H^1(\mathbb{R})}] < \infty, \qquad k = 0, 1. \qquad (3.4.3)$$

Moreover, ψ_0 is the dispersive wave, i.e.,

$$(\psi_0(\cdot, t), \dot{\psi}_0(\cdot, t)) \to 0, \qquad t \to \infty, \qquad (3.4.4)$$

where the convergence holds in metric (3.1.18). Accordingly, it suffices to prove the attraction (3.1.19) only for $Y_1(t) = (\psi_1(\cdot, t), \dot{\psi}_1(\cdot, t))$, i.e.,

$$Y_1(t) \xrightarrow{\mathcal{E}_F} \mathcal{S}, \qquad t \to \pm\infty. \qquad (3.4.5)$$

To prove (3.4.4), we split the initial state $(\psi(x, 0), \dot{\psi}(x, 0))$ into two summands: the first one is smooth with compact support, while the second one has a small energy norm. For solution of the first equation (3.4.2) with smooth initial data, the decay follows by a partial integration in its integral Fourier representation. For small initial data the Sobolev H^1-norm is small uniformly in time by energy conservation.

Furthermore, we extend $\psi_1(x, t)$ and $f(t) := F(\psi(0, t))$ by zero for $t < 0$:

$$\psi_+(x, t) := \begin{cases} \psi_1(x, t), & t > 0, \\ 0, & t < 0, \end{cases} \qquad f_+(t) := \begin{cases} f(t), & t > 0, \\ 0, & t < 0. \end{cases} \qquad (3.4.6)$$

From the second line of formula (3.1.1) it follows that these functions satisfy the equation

$$\ddot{\psi}_+(x, t) = \psi''_+(x, t) - m^2 \psi_+(x, t) + \delta(x) f_+(t), \qquad t > 0 \qquad (3.4.7)$$

in the distribution sense.

Remark 3.4.1 Let us stress that this equation does not coincide with the original Klein–Gordon equation (3.1.1) for $\psi_+(x, t)$, since generally, $f_+(t) \not\equiv F(\psi_+(0, t))$.

The Fourier–Laplace Transform For tempered distributions $g(t)$ of $t \in \mathbb{R}$, we denote by $\tilde{g}(\omega)$ their Fourier transform, which is defined for $g \in C_0^\infty(\mathbb{R})$ by

$$\tilde{g}(\omega) = \int_{\mathbb{R}} e^{i\omega t} g(t)\, dt, \qquad \omega \in \mathbb{R}.$$

The a priori estimates (3.4.3) imply that $\psi_+(x, t)$ and $f_+(t)$ are bounded functions of $t \in \mathbb{R}$ with values in the Sobolev space $H^1(\mathbb{R})$ and in $\mathbb{C}$, respectively. Therefore, their Fourier transforms are (by definition) *quasimeasures* of $\omega \in \mathbb{R}$ with values in $H^1(\mathbb{R})$ and in $\mathbb{C}$, respectively [4].

Let X be a Hilbert space.

Definition 3.4.2 A *quasimeasure* $q(\omega)$ with values in X is the Fourier transform of a function $f(t) \in L^\infty(\mathbb{R}, X)$. The convergence $\tilde{f}_n(\omega) \to \tilde{f}(\omega)$ as $n \to \infty$ in the space of quasimeasures is equivalent to the convergence $f_n(t) \to f(t)$ in $L^\infty_{\text{loc}}(\mathbb{R}, X)$ together with the bound

$$\sup_n \|f_n(t)\|_{L^\infty(\mathbb{R}, X)} < \infty.$$

The Fourier transforms of $\psi_+(x, t)$ and $f_+(t)$ can be extended from the real axis to analytic functions in the upper complex half-plane $\mathbb{C}^+ := \{\omega \in \mathbb{C} : \operatorname{Im} \omega > 0\}$ with values in $H^1(\mathbb{R})$ and in $\mathbb{C}$, respectively:

$$\tilde{\psi}_+(x, \omega) = \int_0^\infty e^{i\omega t} \psi(x, t) \, dt, \qquad \tilde{f}_+(\omega) = \int_0^\infty e^{i\omega t} f(t) \, dt, \qquad \omega \in \mathbb{C}^+.$$

Namely, the functions $e^{-\varepsilon t} \psi_+(x, t)$ and $e^{-\varepsilon t} f_+(t)$ with $\varepsilon \geq 0$ are bounded in $L^\infty(\mathbb{R}, H^1)$ and $L^\infty(\mathbb{R})$ by (3.4.3) and converge in the space $L^\infty_{\text{loc}}(\mathbb{R}, H^1)$ and $L^\infty_{\text{loc}}(\mathbb{R})$, respectively:

$$e^{-\varepsilon t} \psi_+(x, t) \to \psi_+(x, t), \qquad e^{-\varepsilon t} f_+(t) \to f_+(t), \qquad \varepsilon \to 0+.$$

Hence, their Fourier transforms also converge in the sense of quasimeasures by Definition 3.4.2,

$$\tilde{\psi}_+(x, \omega + i\varepsilon) \to \tilde{\psi}_+(x, \omega), \qquad \tilde{f}_+(\omega + i\varepsilon) \to \tilde{f}_+(\omega), \qquad \varepsilon \to 0+. \tag{3.4.8}$$

The limiting absorption principle The Fourier transform of the equation (3.4.7) gives the stationary Helmholtz equation

$$-\omega^2 \tilde{\psi}_+(x, \omega) = \tilde{\psi}''_+(x, \omega) - m^2 \tilde{\psi}_+(x, \omega) + \delta(x) \tilde{f}_+(\omega), \qquad x \in \mathbb{R}. \tag{3.4.9}$$

This equation has a two-parametric family of solutions, but only one of them admits analytic continuation to the upper complex half-plane $\operatorname{Im} \omega > 0$ with values in $H^1(\mathbb{R})$:

$$\tilde{\psi}_+(x, \omega) = -\tilde{f}_+(\omega) \frac{e^{ik(\omega)|x|}}{2ik(\omega)}, \qquad \operatorname{Im} \omega > 0. \tag{3.4.10}$$

Here $k(\omega) := \sqrt{\omega^2 - m^2}$, where the branch has a positive imaginary part for $\operatorname{Im} \omega > 0$. For the other branch, this function *grows exponentially* as $|x| \to \infty$. Such an argument in the selection of solutions of stationary Helmholtz equations is known as the *limiting absorption principle* in diffraction theory [188, 10].

Spectral representation We rewrite (3.4.10) as

$$\tilde{\psi}_+(x,\omega) = \tilde{\alpha}(\omega)e^{ik(\omega)|x|}, \qquad \text{Im } \omega > 0, \qquad \text{where} \quad \alpha(t) := -\psi_+(0,t)$$
$$(3.4.11)$$

It is a nontrivial fact that the identity (3.4.11) between analytic functions keeps its structure for their restrictions to the real axis, which are tempered distributions:

$$\tilde{\psi}_+(x,\omega+i0) = \tilde{\alpha}(\omega+i0)e^{ik(\omega+i0)|x|}, \qquad \omega \in \mathbb{R}, \qquad (3.4.12)$$

where $\tilde{\psi}_+(\cdot, \omega+i0)$ and $\tilde{\alpha}(\omega+i0)$ are the corresponding quasimeasures with values in $H^1(\mathbb{R})$ and $\mathbb{C}$, respectively. The problem is that the factor $M_x(\omega) := e^{ik(\omega+i0)|x|}$ is not smooth with respect to ω at the points $\omega = \pm m$. However, the identity (3.4.12) can be justified by using the theory of quasimeasures [64] (see Lemma B.2 of [64] for details).

The main ideas of the proof are as follows: (i) $M_x^\varepsilon(\omega) := e^{ik(\omega+i\varepsilon)|x|}$ is the *multiplier* in the space of quasimeasures for $\varepsilon \geq 0$, that is, the multiplication by the function $M_x^\varepsilon(\omega)$ is a continuous operator in the convergence of quasimeasures. This follows by the Fourier transform of the product into the convolution and using the fact that $M_x^\varepsilon(\omega) = \tilde{K}_x^\varepsilon(\omega)$, where $K_x^\varepsilon \in L^1(\mathbb{R})$. (ii) $\|K_x^\varepsilon - K_x^0\|_{L^1(\mathbb{R})} \to 0$ as $\varepsilon \to 0+$.

Finally, applying the inverse Fourier transform to (3.4.12), we obtain

$$\psi_+(x,t) = \frac{1}{2\pi}\langle \tilde{\psi}_+(x,\omega+i0), e^{-i\omega t}\rangle$$
$$= \frac{1}{2\pi}\langle \tilde{\alpha}(\omega+i0)e^{ik(\omega+i0)|x|}, e^{-i\omega t}\rangle, \quad x,t \in \mathbb{R}, \qquad (3.4.13)$$

where $\langle \cdot, \cdot \rangle$ is a bilinear duality between distributions with compact support and smooth bounded functions. The right-hand side exists by Theorem 3.5.1; see below.

3.5 A Nonlinear Analog of Kato's Theorem

It turns out that the properties of the quasimeasure $\tilde{\alpha}(\omega+i0)$ for $|\omega| < m$ and that for $|\omega| > m$ differ significantly. This is because the set $\{i\omega : |\omega| \geq m\}$ is the continuous spectrum of the operator

$$A = \begin{pmatrix} 0 & 1 \\ \frac{d^2}{dx^2} - m^2 & 0 \end{pmatrix},$$

which is the generator of the linear part of (3.1.5). The following theorem plays a key role in the proof of Theorem 3.1.2. Denote $\Sigma := \{\omega \in \mathbb{R} : |\omega| > m\}$. Below we will also write $\tilde{\alpha}(\omega)$ and $k(\omega)$ instead of $\tilde{\alpha}(\omega + i0)$ and $k(\omega + i0)$ for $\omega \in \mathbb{R}$.

Theorem 3.5.1 ([64, Proposition 3.2]). *Let conditions (3.1.3), (3.1.6), and (3.1.8) hold, and let $\psi(t) \in C(\mathbb{R}, \mathcal{E})$ be any finite-energy solution of (3.1.5). Then the corresponding tempered distribution $\tilde{\alpha}(\omega)$ is absolutely continuous on Σ. Moreover, $\alpha \in L^1(\Sigma)$ and*

$$\int_{\Sigma} |\tilde{\alpha}(\omega)|^2 |\omega k(\omega)| \, d\omega < \infty. \tag{3.5.1}$$

Proof We first explain the main idea of the proof. By (3.4.13), the function $\psi_+(x,t)$ is formally a "linear combination" of the functions $e^{ik|x|}$ with the amplitudes $\hat{z}(\omega)$:

$$\psi_+(x,t) = \frac{1}{2\pi} \int_{\mathbb{R}} \hat{z}(\omega) e^{ik(\omega)|x|} e^{-i\omega t} \, d\omega, \qquad x \in \mathbb{R}.$$

For $\omega \in \Sigma$ the functions $e^{ik(\omega)|x|}$ have an infinite $L^2(\mathbb{R})$-norm, whereas $\psi_+(\cdot,t)$ has a finite $L^2(\mathbb{R})$-norm. This is possible only if the amplitude is absolutely continuous in Σ. This idea is suggested by the Fourier integral $f(x) = \int_{\mathbb{R}} e^{-ikx} g(k) dk$, which belongs to $L^2(\mathbb{R})$ if and only if $g \in L^2(\mathbb{R})$. For example, if one took $\hat{z}(\omega) = \delta(\omega - \omega_0)$ with $\omega_0 \in \Sigma$, then $\psi_+(\cdot,t)$ will be of infinite L^2-norm.

The rigorous proof relies on estimates of Paley–Wiener type. First note that the Parseval–Plancherel identity and (3.4.3) imply

$$\int_{\mathbb{R}} \|\tilde{\psi}_+(\cdot,\omega + i\varepsilon)\|^2_{H^1(\mathbb{R})} \, d\omega = 2\pi \int_0^{\infty} e^{-2\varepsilon t} \|\psi_+(\cdot,t)\|^2_{H^1(\mathbb{R})} \, dt$$

$$\leq \frac{\text{const}}{\varepsilon}, \qquad \varepsilon > 0. \tag{3.5.2}$$

On the other hand, we can exactly estimate the integral on the left-hand side of (3.5.2). Indeed, according to (3.4.13),

$$\tilde{\psi}_+(\cdot,\omega + i\varepsilon) = \tilde{\alpha}(\omega + i\varepsilon) e^{ik(\omega+i\varepsilon)|x|}.$$

Consequently, (3.5.2) gives us that

$$\varepsilon \int_{\mathbb{R}} |\tilde{\alpha}(\omega + i\varepsilon)|^2 \|e^{ik(\omega+i\varepsilon)|x|}\|^2_{H^1(\mathbb{R})} \, d\omega \leq \text{const}, \qquad \varepsilon > 0. \tag{3.5.3}$$

Here is a crucial observation about the asymptotics of the norm of $e^{ik(\omega+i\varepsilon)|x|}$ as $\varepsilon \to 0+$.

Lemma 3.5.2 *(i) For $\omega \in \mathbb{R}$,*

$$\lim_{\varepsilon \to 0+} \varepsilon \|e^{ik(\omega+i\varepsilon)|x|}\|^2_{H^1(\mathbb{R})} = n(\omega) := \begin{cases} \omega k(\omega), & |\omega| > m \\ 0, & |\omega| < m \end{cases}, \qquad (3.5.4)$$

where the norm in $H^1(\mathbb{R})$ is chosen to be $\|\psi_{H^1(\mathbb{R})}\| = \left(\|\psi'_{L^2}\|^2 + m^2\|\psi\|^2_{L^2(\mathbb{R})}\right)^{1/2}$.

(ii) For any $\delta > 0$, there exists an $\varepsilon_\delta > 0$ such that for $|\omega| > m + \delta$ and $\varepsilon \in (0, \varepsilon_\delta)$,

$$\varepsilon \|e^{ik(\omega+i\varepsilon)|x|}\|^2_{H^1(\mathbb{R})} \geq \frac{n(\omega)}{2}. \qquad (3.5.5)$$

Proof Let us compute the $H^1(\mathbb{R})$-norm using the Fourier space representation. Let us set $k_\varepsilon = k(\omega + i\varepsilon)$ so that Im $k_\varepsilon > 0$; we then obtain $F_{x \to k}\left[e^{ik_\varepsilon|x|}\right] = 2ik_\varepsilon/(k_\varepsilon^2 - k^2)$ for $k \in \mathbb{R}$. Hence, by the Cauchy theorem on residues, we have

$$\|e^{ik_\varepsilon|x|}\|^2_{H^1(\mathbb{R})} = \frac{2|k_\varepsilon|^2}{\pi} \int_{\mathbb{R}} \frac{(k^2 + m^2)dk}{|k_\varepsilon^2 - k^2|^2} = -4\,\mathrm{Im}\left[\frac{(k_\varepsilon^2 + m^2)\bar{k}_\varepsilon}{k_\varepsilon^2 - \bar{k}_\varepsilon^2}\right]. \qquad (3.5.6)$$

Substituting here $k_\varepsilon^2 = (\omega + i\varepsilon)^2 - m^2$, we see that

$$\|e^{ik(\omega+i\varepsilon)|x|}\|^2_{H^1(\mathbb{R})} = \frac{1}{\varepsilon}\mathrm{Re}\left[\frac{(\omega + i\varepsilon)^2\overline{k(\omega + i\varepsilon)}}{\omega}\right], \quad \varepsilon > 0, \ \omega \in \mathbb{R}, \ \omega \neq 0.$$

The limits (3.5.4) now follow, since the function $k(\omega)$ is real for $|\omega| > m$ but is purely imaginary for $|\omega| < m$. Therefore, the second assertion of the lemma also follows, since $n(\omega) > 0$ for $|\omega| > m$, and $n(\omega) \sim |\omega|^2$ for $|\omega| \to \infty$. $\qquad \square$

Remark 3.5.3 Clearly, $n(\omega) \equiv 0$ for $|\omega| < m$ without any calculations, since in that case the function $e^{ik(\omega)|x|}$ decays exponentially in x, and so the $H^1(\mathbb{R})$-norm of $e^{ik(\omega+i\varepsilon)|x|}$ remains finite as $\varepsilon \to 0+$.

Substituting (3.5.5) into (3.5.3), we obtain

$$\int_{\Sigma_\delta} |\tilde{\alpha}(\omega + i\varepsilon)|^2 \omega k(\omega)\, d\omega \leq 2C, \qquad 0 < \varepsilon < \varepsilon_\delta, \qquad (3.5.7)$$

with the same C as in (3.5.3), and with the region $\Sigma_\delta := \{\omega \in \mathbb{R} : |\omega| > m + \delta\}$. We conclude that for each $\delta > 0$ the set of functions

$$g_\varepsilon(\omega) = \tilde{\alpha}(\omega + i\varepsilon)|\omega k(\omega)|^{1/2}, \qquad \varepsilon \in (0, \varepsilon_\delta),$$

is bounded in the Hilbert space $L^2(\Sigma_\delta)$, so it is weakly compact by the Banach theorem. Hence, convergence of the distributions (3.4.8) implies weak convergence in $L^2(\Sigma_\delta)$:

$$g_\varepsilon \rightharpoonup g, \qquad \varepsilon \to 0+,$$

where the limit function $g(\omega)$ coincides with the distribution $\hat{z}(\omega)|\omega k(\omega)|^{1/2}$ restricted to Σ_δ. It remains to note that the norms of g in $L^2(\Sigma_\delta)$ for all $\delta > 0$ are bounded in view of (3.5.7), and this implies (3.5.1). Finally, $\tilde{\alpha}(\omega) \in L^1(\bar{\Sigma})$ by (3.5.1) and by the Cauchy–Schwarz inequality.   □

Remark 3.5.4 Theorem 3.5.1 is a nonlinear analog of Kato's theorem on the absence of embedded eigenvalues in the continuous spectrum. Indeed, solutions of type $\psi_*(x)e^{-i\omega_* t}$ become $\psi_*(x)[\pi i \delta(\omega - \omega_*) + \text{v.p.}\frac{1}{i(\omega - \omega_*)}]$ in the Fourier–Laplace transform, and this is forbidden for $|\omega_*| > m$ by Theorem 3.5.1.

3.6 Splitting into Dispersive and Bound Components

Theorem 3.5.1 presupposes a splitting of the solutions (3.4.13) into a "dispersive component" and a "bound component":

$$\psi_+(x,t) = \frac{1}{2\pi} \int\limits_\Sigma (1 - \zeta(\omega))\tilde{\alpha}(\omega)e^{ik(\omega)|x|}e^{-i\omega t} d\omega$$

$$+ \frac{1}{2\pi}\langle \zeta(\omega)\tilde{\alpha}(\omega)e^{ik(\omega)|x|}, e^{-i\omega t}\rangle$$

$$= \psi_d(x,t) + \psi_b(x,t), \qquad x \in \mathbb{R}, \quad t \in \mathbb{R}, \qquad (3.6.1)$$

where

$$\zeta(\omega) \in C_0^\infty(\mathbb{R}), \quad \text{and} \quad \zeta(\omega) = 1 \quad \text{for} \quad \omega \in [-m-1, m+1].$$

Note that $\psi_d(x,t)$ is a dispersive wave, because

$$\psi_d(x,t) := \frac{1}{2\pi} \int\limits_\Sigma (1 - \zeta(\omega))e^{-i\omega t}\tilde{\alpha}(\omega)e^{ik(\omega)|x|} d\omega \to 0, \qquad t \to \infty$$

$$(3.6.2)$$

according to the Riemann–Lebesgue theorem, since $\alpha \in L^1(\Sigma)$ by Theorem 3.5.1. Moreover, it is easy to prove that

$$(\psi_d(\cdot,t), \dot{\psi}_d(\cdot,t)) \to 0, \qquad t \to \infty \qquad (3.6.3)$$

in the metric (3.1.18). The main ideas of the proof are as follows: (3.5.1) implies that

$$\psi_d(\cdot,t), \dot{\psi}_d(\cdot,t), \nabla\psi_d(\cdot,t) \in L^2(\mathbb{R}), \qquad (3.6.4)$$

which follows by splitting of (3.6.2) in two integrals over $(-\infty, -m-1]$ and over $[m+1, \infty)$ and applying the Parseval–Plancherel theorem to each integral in the variable $k = k(\omega)$. Finally, the function $\tilde{\alpha}(\omega)$ admits the splitting into two summands $\tilde{\alpha}(\omega) = \tilde{\alpha}_1(\omega) + \tilde{\alpha}_2(\omega)$, where $\tilde{\alpha}_1(\omega) \in C_0^\infty(\mathbb{R})$, while the integral of type (3.5.1) for $\tilde{\alpha}_2(\omega)$ is small. Then integration by parts in the integrals with $\tilde{\alpha}_1(\omega)$ implies (3.6.4).

The decay (3.6.4) means that it remains to prove the attraction (3.4.5) with the function $Y_b(t) := (\psi_b(\cdot,t), \dot{\psi}_b(\cdot,t))$ replacing $Y_1(t)$:

$$Y_b(t) \xrightarrow{\mathcal{E}_F} \mathcal{S}, \qquad t \to \infty. \qquad (3.6.5)$$

3.7 Omega-Compactness

Here we establish the omega-compactness of the trajectory $Y_b(t)$, which is necessary for the application of Lemma 3.3.3. First, note that the bound component $\psi_b(x,t)$ is a smooth function for $x \neq 0$, and

$$\partial_x^n \partial_t^l \psi_b(x,t)$$
$$= \frac{1}{2\pi} \langle \zeta(\omega)(ik(\omega)\operatorname{sgn}x)^n \tilde{\alpha}(\omega)e^{ik(\omega)|x|}, (-i\omega)^l e^{-i\omega t}\rangle, \quad x \neq 0, \quad t \in \mathbb{R}$$
$$(3.7.1)$$

for any $n, l = 0, 1, \ldots$. These formulas must be justified, since the function $k(\omega)$ is not smooth at the points $\omega = \pm m$. The case $n = 0$ is trivial. For $n \geq 1$ the justification is done in [62, 64] by a suitable development of the theory of quasimeasures [4]. The main idea is as follows: for $x \neq 0$ we have

$$\zeta(\omega)\frac{e^{ik(\omega)|x+\varepsilon|} - e^{ik(\omega)|x|}}{\varepsilon} = \tilde{f}_x^\varepsilon(\omega), \qquad (3.7.2)$$

where $f_x^\varepsilon(\cdot) \in L^1(\mathbb{R})$ and $f_x^\varepsilon(\cdot) \to F_{\omega \to t}^{-1}\zeta(\omega)ik(\omega)\operatorname{sgn}x\, e^{ik(\omega)|x|}$ as $\varepsilon \to 0$, where the convergence holds in $L^1(\mathbb{R})$. Further arguments follow the same line as the justification of (3.4.12). See also Proposition 4.1 of [64]. These formulas imply the boundedness of each derivative.

Lemma 3.7.1 ([64, Proposition 4.1]). *For all* $j, l = 0, 1, 2, \ldots$.

$$\sup_{x \neq 0} \sup_{t \in \mathbb{R}} |\partial_x^n \partial_t^l \psi_b(x,t)| < \infty. \qquad (3.7.3)$$

Proof Note that in general the distribution $\tilde{\alpha}(\omega)$ is not a finite measure, since we only know that $\alpha(t) := \psi_+(0,t)$ is a bounded function by (3.4.11) and (3.1.7). To prove the lemma, it suffices to check that

$$\zeta(\omega)(ik(\omega)\,\text{sgn}\ x)^n e^{ik(\omega)|x|}(-i\omega)^l = \tilde{g}_x(\omega), \qquad (3.7.4)$$

where the function $g_x(\cdot)$ belongs to a bounded subset of $L^1(\mathbb{R})$ for $x \neq 0$ and $t \in \mathbb{R}$. This implies the lemma, since by the Parseval–Plancherel identity the right-hand side of (3.7.1) is the convolution

$$\langle \alpha(t-s), g_x(s) \rangle,$$

where $\alpha(t)$ is a bounded function. $\qquad\square$

Remark 3.7.2 All properties of quasimeasures used above and below are justified in [62, 64] by similar arguments relying on the Parseval–Plancherel identity.

By the Ascoli–Arzelà theorem, the estimates (3.7.3) imply that for any sequence $s_j \to \infty$, there is a subsequence $s_{j'} \to \infty$ such that

$$\partial_x^n \partial_t^l \psi_b(x, s_{j'} + t) \to \partial_x^n \partial_t^l \beta(x,t), \qquad x \neq 0,\ t \in \mathbb{R} \qquad (3.7.5)$$

for any $n, l = 0, \ldots$. Moreover, this convergence is uniform for $|x| + |t| \leq R$ with any $R > 0$, and

$$\sup_{x \neq 0} \sup_{t \in \mathbb{R}} |\partial_x^n \partial_t^l \beta(x,t)| < \infty. \qquad (3.7.6)$$

Corollary 3.7.3 *Each solution $Y(t) \in C(\mathbb{R}, \mathcal{E})$ to (3.1.5) is omega-compact. This follows from (3.4.4), (3.6.3), and (3.7.5).*

3.8 Reduction of Spectrum to Spectral Gap

For $n = l = 0$ the convergence (3.7.5) and the representation (3.7.1) imply the convergence of their Fourier transforms: for each $x \in \mathbb{R}$,

$$\zeta(\omega)\tilde{\alpha}(\omega)e^{ik(\omega)|x|}e^{-i\omega s_{j'}} \to \tilde{\beta}(x,\omega), \qquad j' \to \infty \qquad (3.8.1)$$

in the sense of quasimeasures according to Definition 3.4.2.

Lemma 3.8.1 *For any $x \in \mathbb{R}$,*

$$\tilde{\beta}(x,\omega) = 0, \qquad |\omega| > m. \qquad (3.8.2)$$

Proof The convergence (3.8.1) implies that for each $x \in \mathbb{R}$,

$$\zeta(\omega)\tilde{\alpha}(\omega)e^{-i\omega s_{j'}} \to \tilde{\gamma}(\omega) := \tilde{\beta}(x,\omega)e^{-ik(\omega)|x|}, \qquad j' \to \infty \qquad (3.8.3)$$

in the sense of quasimeasures, since $e^{-ik(\omega)|x|}$ is a multiplier. For the same reason, we obtain $\tilde{\beta}(x,\omega) = \tilde{\gamma}(\omega)e^{ik(\omega)|x|}$, and hence,

$$\beta(x,t) = \frac{1}{2\pi}\langle\tilde{\gamma}(\omega)e^{ik(\omega)|x|}, e^{-i\omega t}\rangle, \qquad (x,t) \in \mathbb{R}^2. \qquad (3.8.4)$$

Note that

$$\beta(0,t) = \gamma(t). \qquad (3.8.5)$$

Finally, the key observation is that (3.8.3) and Theorem 3.5.1 imply that

$$\operatorname{supp} \tilde{\gamma} \subset [-m,m] \qquad (3.8.6)$$

by the Riemann–Lebesgue theorem. $\qquad\square$

Remark 3.8.2 Note that the convergence (3.8.1) in the sense of tempered distributions is not enough for the proof of Lemma 3.8.1. It is important that the function $e^{-ik(\omega)|x|}$ is a multiplier in the space of quasimeasures.

3.9 Reduction of Spectrum to a Single Point

The question arises of the available means for verifying the representation (3.3.2) for omega-limit trajectories. We have no formulas for solutions of the nonlinear equation (3.1.1), and so the only hope is to use the nonlinear equation itself.

Equation for Omega-Limit Trajectories and Spectral Inclusion

The key observation, albeit simple, is that $\beta(x,t)$ is a solution of nonlinear equation (3.1.1) *for all* $t \in \mathbb{R}$.

Lemma 3.9.1 *The function $\beta(x,t)$ satisfies the original equation (3.1.5):*

$$\ddot{\beta}(x,t) = \beta'(x,t) - m^2\beta(x,t) + \delta(x)F(\beta(0,t)), \qquad (x,t) \in \mathbb{R}^2. \qquad (3.9.1)$$

Proof This lemma follows by (3.4.4), (3.6.3), and (3.7.5) in the limit as $s_{j'} \to \infty$ in the equation (3.4.7), where

$$\psi_+(x,s_{j'}+t) = \psi_d(x,s_{j'}+t) + \psi_b(x,s_{j'}+t),$$
$$f_+(t) = F(\psi_0(0,s_{j'}+t) + \psi_d(0,s_{j'}+t) + \psi_b(0,s_{j'}+t)). \qquad \square$$

Remark 3.9.2 (i) Let us recall that the function $\psi_+(x, s_{j'} + t)$ does not satisfy the original equation (3.1.1) by Remark 3.4.1.

(ii) The limit equation (3.9.1) holds for all $t \in \mathbb{R}$, whereas the equation (3.4.7) holds only for $t > 0$.

Applying the Fourier transform to equation (3.9.1), we now get the corresponding *nonlinear stationary Helmholtz equation*

$$-\omega^2 \tilde{\beta}(x,\omega) = \tilde{\beta}''(x,\omega) - m^2 \tilde{\beta}(x,\omega) + \delta(x)\tilde{f}(\omega), \qquad (x,\omega) \in \mathbb{R}^2, \quad (3.9.2)$$

where we define $f(t) := F(\beta(0,t)) = F(\gamma(t))$ in accordance with (3.8.5). From (3.1.9), we see that

$$f(t) = a(|\gamma(t)|)\gamma(t) = A(t)\gamma(t), \qquad A(t) := a(|\gamma(t)|), \qquad t \in \mathbb{R}.$$

Furthermore, applying the Fourier transform, we obtain the convolution $\tilde{f} = \tilde{A} * \tilde{\gamma}$, which exists by (3.8.6). Respectively, (3.9.2) is now

$$-\omega^2 \tilde{\beta}(x,\omega) = \tilde{\beta}''(x,\omega) - m^2 \tilde{\beta}(x,\omega) + \delta(x)[\tilde{A} * \tilde{\gamma}](\omega), \qquad (x,\omega) \in \mathbb{R}^2.$$

This identity implies the key **spectral inclusion**

$$\text{supp } \tilde{A} * \tilde{\gamma} \subset \text{supp } \tilde{\gamma}, \qquad (3.9.3)$$

because

$$\text{supp } \tilde{\beta}(x,\cdot) \subset \text{supp } \tilde{\gamma}, \qquad \text{supp } \tilde{\beta}''(x,\cdot) \subset \text{supp } \tilde{\gamma} \qquad (3.9.4)$$

in view of the representation (3.8.4). In the next section we will derive (3.3.2) from the inclusion (3.9.3), using a fundamental result of harmonic analysis – the Titchmarsh convolution theorem.

Titchmarsh Convolution Theorem
In 1926, E.C. Titchmarsh proved a theorem on the distribution of zeros of entire functions (see [79] and [73, p. 119]), which has, in particular, the following corollary (see [6, Theorem 4.3.3]):

Theorem. *Let $f(\omega)$ and $g(\omega)$ be distributions of $\omega \in \mathbb{R}$ with bounded supports. Then*

$$[\text{supp } f * g] = [\text{supp } f] + [\text{supp } g],$$

*where $[X]$ denotes the **convex hull** of the set $X \subset \mathbb{R}$.*

Proof of Theorem 3.1.2 By Lemma 3.3.3 and Corollary 3.7.3, it suffices to check the representation (3.3.2) for the above constructed omega-limit trajectory $\beta(x,t)$, since every omega-limit trajectory can be obtained in this way by (3.4.4), (3.6.3), and (3.7.3).

First, we know that supp $\tilde{\gamma}$ is bounded because of (3.8.6). Consequently, supp $\tilde{A}$ is also bounded, since $A(t) := a(|\gamma(t)|)$ is a polynomial in $|\gamma(t)|^2$ according to (3.1.16). Now the spectral inclusion (3.9.3) and the Titchmarsh theorem imply

$$[\text{supp } \tilde{A}] + [\text{supp } \tilde{\gamma}] \subset [\text{supp } \tilde{\gamma}],$$

whence it immediately follows that $[\text{supp } \tilde{A}] = \{0\}$. Besides, $A(t) := a(|\gamma(t)|)$ is a bounded function due to (3.7.6), because $\gamma(t) = \beta(0,t)$. Therefore, $\tilde{A}(\omega) = C\delta(\omega)$, and hence

$$a(|\gamma(t)|) = C_1, \qquad t \in \mathbb{R}. \tag{3.9.5}$$

Now, the strict nonlinearity condition (3.1.16) implies that

$$|\gamma(t)| = C_2, \qquad t \in \mathbb{R}. \tag{3.9.6}$$

This immediately gives

$$\text{supp } \tilde{\gamma} = \{\omega_+\} \tag{3.9.7}$$

by the same Titchmarsh theorem for the convolution $\tilde{\gamma} * \overline{\tilde{\gamma}} = C_3\delta(\omega)$. Therefore, we have $\tilde{\gamma}(\omega) = C_4\delta(\omega - \omega_+)$, and now (3.3.2) follows from (3.8.4). Theorem 3.1.2 is proved.

Remark 3.9.3 In the case of the Schrödinger equation (3.2.4), the Titchmarsh theorem does not work. The fact is that the continuous spectrum of the operator $-d^2/dx^2$ is the half-line $[0,\infty)$, so now the role of the spectral gap is played by the unbounded interval $(-\infty,0)$. Accordingly, in this case the spectral inclusion (3.10.1) gives only that supp $\tilde{\beta}(x,\cdot) \subset (-\infty,0)$, while the Titchmarsh theorem applies only to distributions with bounded supports.

3.10 On the Nonlinear Radiative Mechanism

Let us explain the informal arguments for the attraction to stationary orbits behind the formal proof of Theorem 3.1.2. The main part of the proof involves the study of the spectrum of omega-limit trajectories

$$\beta(x,t) = \lim_{s_{j'} \to \infty} \psi(x, s_{j'} + t).$$

Theorem 3.5.1 implies the spectral inclusion (3.8.6), which leads to

$$\text{supp } \tilde{\beta}(x,\cdot) \subset [-m,m], \qquad x \in \mathbb{R}. \tag{3.10.1}$$

The Titchmarsh theorem then allows us to conclude that

$$\text{supp } \tilde{\beta}(x,\cdot) = \{\omega_+\}. \tag{3.10.2}$$

These two inclusions are suggested by the following two informal arguments.

A. *Dispersive radiation in the continuous spectrum.*

B. *Nonlinear spreading of spectrum and energy transfer from lower to higher harmonics.*

A. Dispersive radiation in the continuous spectrum. The inclusion (3.10.1) is due to the dispersive mechanism, which can be illustrated by the example of energy radiation in a wave field of a harmonic source with frequency lying in the continuous spectrum. Namely, let us consider a 1D linear Klein–Gordon equation with a *harmonic source*

$$\ddot{\psi}(x,t) = \psi''(x,t) - m^2\psi(x,t) + b(x)e^{-i\omega_0 t}, \qquad x \in \mathbb{R}, \tag{3.10.3}$$

where the amplitude $b \in L^2(\mathbb{R})$ and the real frequency ω_0 is different from $\pm m$. In this case, the *limiting amplitude principle* holds [188, 72, 75]:

$$\psi(x,t) \sim a(x)e^{-i\omega_0 t}, \qquad t \to \infty. \tag{3.10.4}$$

For equation (3.10.3), this follows directly from the Fourier–Laplace transform in time

$$\tilde{\psi}(\omega,t) = \int_0^\infty e^{i\omega t}\psi(x,t)dt, \qquad x \in \mathbb{R}, \quad \text{Im } \omega > 0. \tag{3.10.5}$$

Indeed, applying this transform to equation (3.10.3), we see that

$$-\omega^2\tilde{\psi}(x,\omega) = \tilde{\psi}''(x,\omega) - m^2\tilde{\psi}(x,\omega) + \frac{b(x)}{i(\omega - \omega_0)}, \qquad x \in \mathbb{R}, \quad \text{Im } \omega > 0,$$

where for simplicity we assume zero initial data. Hence,

$$\tilde{\psi}(\cdot,\omega) = \frac{R(\omega)b}{i(\omega - \omega_0)} = \frac{R(\omega_0 + i0)b}{i(\omega - \omega_0)} + \frac{R(\omega)b - R(\omega_0 + i0)b}{i(\omega - \omega_0)}, \qquad \text{Im } \omega > 0, \tag{3.10.6}$$

where

$$R(\omega) := (H - \omega^2)^{-1}$$

is the resolvent of the Schrödinger operator $H := -d^2/dx^2 + m^2$. This resolvent is a convolution operator with the fundamental solution $-\frac{e^{ik(\omega)|x|}}{2ik(\omega)}$, where $k(\omega) = \sqrt{\omega^2 - m^2} \in \overline{\mathbb{C}^+}$ for $\omega \in \overline{\mathbb{C}^+}$, as in (3.4.10). The last fraction of (3.10.6) is regular at $\omega = \omega_0$, and therefore its contribution is a dispersive wave, which decays like (3.6.3) in local energy seminorms. Consequently, the long-time asymptotics of $\psi(x,t)$ are determined by the middle quotient in (3.10.6). Therefore, (3.10.4) holds with the limiting amplitude $a(x) = R(\omega_0 + i0)b$. The Fourier transform of this limiting amplitude is equal to

$$\hat{a}(k) = -\frac{\hat{b}(k)}{k^2 + m^2 - (\omega_0 + i0)^2}, \qquad k \in \mathbb{R}.$$

This formula shows that the properties of the limiting amplitudes differ significantly in the cases $|\omega_0| < m$ and $|\omega_0| \geq m$: $a(x) \in H^2(\mathbb{R})$ for $|\omega_0| < m$; however,

$$a(x) \notin L^2(\mathbb{R}) \quad \text{for} \quad |\omega_0| \geq m, \tag{3.10.7}$$

if $|\hat{b}(k)| > 0$ in a neighborhood of the "sphere" $|k|^2 + m^2 = \omega_0^2$ (which consists of two points in the 1D case). *This means the following.*

I. In the case $|\omega_0| \geq m$ the energy of the solution $\psi(x,t)$ tends to infinity for large times according to (3.10.4) and (3.10.7). This means that energy is transmitted from the harmonic source to the wave field!

II. In the opposite case $|\omega_0| < m$ the energy of the solution remains bounded, so there is no energy radiation.

It is exactly this energy radiation in the case of $|\omega_0| \geq m$ that prohibits the occurrence of harmonics with such frequencies in omega-limit trajectories. Indeed, any omega-limit trajectory cannot radiate at all, since the total energy is finite and bounded from below, and hence the radiation cannot last forever. These physical arguments make the inclusion (3.10.1) plausible, although a rigorous proof of it, as was seen above, requires special arguments.

Recall that the set $i\Sigma := \{i\omega_0 : \omega_o \in \mathbb{R}, |\omega_0| \geq m\}$ coincides with the continuous spectrum of the generator of the free Klein–Gordon equation. Radiation in the continuous spectrum is well known in the theory of waveguides. Namely, waveguides can transmit only signals with a frequency $|\omega_0| > \mu$, where μ is a *threshold frequency*, which is an edge point of the continuous spectrum [74]. In our case, the waveguide occupies the "entire space" $x \in \mathbb{R}$ and is described by the nonlinear Klein–Gordon equation (3.1.1) with the threshold frequency m.

B. Nonlinear inflation of spectrum and energy transfer from lower to higher harmonics. Let us show that the single-frequency spectrum (3.10.2)

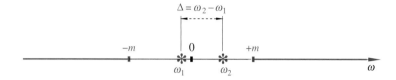

Figure 3.1 Two-point spectrum.

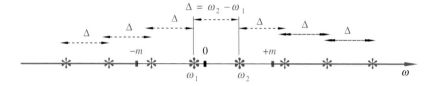

Figure 3.2 Nonlinear inflation of spectrum.

is due to the inflation of the spectrum by nonlinear functions. For example, consider the potential $U(\psi) = |\psi|^4$. Accordingly, $F(\psi) = -\nabla_{\overline{\psi}} U(\psi) = -4|\psi|^2\psi$. We consider the sum $\psi(t) = e^{i\omega_1 t} + e^{i\omega_2 t}$ of two harmonics, whose spectrum is shown in Figure 3.1.

Let us substitute this sum into the nonlinearity:

$$F(\psi(t)) \sim \psi(t)\overline{\psi(t)}\psi(t) = e^{i\omega_2 t} e^{-i\omega_1 t} e^{i\omega_2 t} + \dots$$
$$= e^{i(\omega_2 + \Delta)t} + \dots, \qquad \Delta := \omega_2 - \omega_1.$$

The spectrum of this expression contains harmonics with the new frequency $\omega_2 + \Delta$. As a result, all the frequencies $\omega_1 - \Delta$, $\omega_1 - 2\Delta$, $\dots$ and $\omega_2 + \Delta$, $\omega_2 + 2\Delta$, $\dots$ also will appear in the nonlinear dynamics described by (3.1.1) (see Figure 3.2). Consequently, these frequencies will appear also in the nonlinear δ-function term which plays the role of a source.

As we already know, these frequencies, which lie in the continuous spectrum $|\omega| > m$, will surely cause energy radiation. This radiation will continue until the spectrum of the solution contains at least two different frequencies. It is this fact that prohibits the presence of two different frequencies in omega-limit trajectories, because the total energy is finite, and thus the radiation cannot continue forever.

However, we stress that

(i) the precise meaning of the arguments "until the spectrum of the solution contains at least two different frequencies" is given by the concept of omega-limit trajectory;

(ii) the inflation of the spectrum by a nonlinearity is justified by the Titchmarsh convolution theorem.

Nonlinear radiative mechanism The above arguments physically mean that the following binary *nonlinear radiative mechanism* takes place:

I. a nonlinearity inflates the spectrum, which means energy transfer from lower to higher harmonics;

II. the dispersive radiation carries energy to infinity.

For the first time we have rigorously justified such a nonlinear radiative mechanism for the nonlinear $U(1)$-invariant Klein–Gordon and Dirac equations (3.1.5) and (3.2.1)–(3.2.3), and for other models, see [50, 60, 61, 65, 66, 67, 69, 70]. Our numerical experiments demonstrate a similar nonlinear radiative mechanism for *relativistic nonlinear wave equations* (see Remark 6.1.1). However, a rigorous proof is still missing.

3.11 Conjecture on Attractors of *G*-Invariant PDEs

Let us specify the conjecture (6) for *generic* Hamiltonian *G-invariant* PDEs in $\mathbb{R}^n$ of type (5) with a Lie symmetry group G acting on suitable Hilbert or Banach phase space $\mathcal{E}$ via a linear representation T. The Hamiltonian structure means that

$$F(\Psi) = JD\mathcal{H}(\Psi), \qquad J^* = -J, \tag{3.11.1}$$

where $\mathcal{H}$ denotes the corresponding Hamiltonian functional. The G-invariance means that

$$F(T(g)\Psi) = T(g)F(\Psi), \qquad \Psi \in \mathcal{E} \tag{3.11.2}$$

for all $g \in G$. In that case, for any solution $\Psi(t)$ to equations (5), the trajectory $T(g)\Psi(t)$ is also a solution, so the representation commutes with the dynamical group $U(t)\colon \Psi(0) \mapsto \Psi(t)$:

$$T(g)U(t) = U(t)T(g). \tag{3.11.3}$$

Let us note that the theory of elementary particles deals systematically with the symmetry groups $SU(2)$, $SU(3)$, $SU(5)$, $SO(10)$, and others, as well as with the group

$$SU(4) \times SU(2) \times SU(2),$$

which is the symmetry group of "Grand Unification"; see [223].

The conjecture (6) means that all solutions of type $e^{\hat{\lambda}t}\Psi$ with $\lambda \in \mathfrak{g}$ and $\Psi \in \mathcal{E}$ form a global attractor for **generic** G-invariant Hamiltonian nonlinear PDEs of type (5).

We must still specify the meaning of the term **generic** G**-invariant equation** in Conjecture (6) (and in all results of Chapters 1, 2, and 3). Namely, this conjecture means that the asymptotics (6) hold for all solutions for an *open dense set* of G-invariant equations.

In particular, all asymptotics (7), (12), (15), and (16) hold under appropriate conditions, which define some "open dense subset" of G-invariant equations with the four types of the symmetry group G. These asymptotic expressions may break down if these conditions fail – this corresponds to some "exceptional" equations. For example, global attraction (3.1.2) breaks down for the linear Schrödinger equations with at least two different eigenvalues. Thus, linear equations are exceptional, **not generic**!

The general situation is the following. Let a Lie group G_1 be a (proper) subgroup of some larger Lie group G_2. Then G_2-invariant equations form an "exceptional subset" among all G_1-invariant equations, and the corresponding asymptotics (6) may be completely different. For example, the trivial group $\{e\}$ is a subgroup in $U(1)$ and in $\mathbb{R}^n$, and the asymptotic expressions (12) and (15) may differ significantly from (7).

Conjecture (6) is confirmed by all rigorous results [40]–[71] presented in previous sections of this book. The results concern a list of model equations of type (5) with the following four basic symmetry groups: the trivial group $\{e\}$, the group of translations $\mathbb{R}^n$, the unitary group $U(1)$, and the orthogonal group $SO(3)$. In these cases, the asymptotics (6) read as (7), (12), (15), and (16), respectively.

Conjecture (6) suggests to define *stationary G-orbits* for equations (5) as solutions of type

$$\Psi(t) = e^{\hat{\lambda}t}\Psi, \qquad t \in \mathbb{R}, \tag{3.11.4}$$

where $\lambda \in \mathfrak{g}$. This definition leads to the corresponding *nonlinear eigenvalue problem*

$$F(\Psi) = \hat{\lambda}\Psi. \tag{3.11.5}$$

In particular, for the case of unitary symmetry group $U(1)$, the Lie algebra is $\mathfrak{g} = \mathbb{R}$, and λ is a real number. On the other hand, for the symmetry group $G = SU(3)$, the generator λ is a skew-Hermitian 3×3-matrix.

Empirical evidence Conjecture (6) agrees with the Gell-Mann–Ne'eman theory of baryons [222, 224]. Namely, in 1961 Gell-Mann and Ne'eman suggested using the symmetry group $SU(3)$ for the strong interaction of baryons relying on the discovered parallelism between empirical data for the baryons, and the "Dynkin scheme" of the Lie algebra $\mathfrak{g} = su(3)$ with 8 generators (the famous "eightfold way").

This theory resulted in the scheme of quarks in quantum chromodynamics [223] and in the prediction of a new baryon with prescribed values of its mass and decay products. This particle (the Ω^--hyperon) was promptly discovered experimentally [221].

The elementary particles seem to describe long-time asymptotics of quantum fields. Hence, this empirical correspondence between elementary particles and generators of the Lie algebras presumably gives evidence in favor of our general conjecture on attractors (6) for equations with Lie symmetry groups.

4

Asymptotic Stability of Stationary Orbits
and Solitons

This chapter concerns "asymptotic stability of solitary manifolds," which means a local attraction, i.e., for states sufficiently close to such manifolds. In Sections 4.1 and 4.2 we describe general strategies introduced by A. Soffer and M. Weinstein and by V. S. Buslaev and G. Perelman for proving such local attraction. In Sections 4.3 and 4.4 we give a brief survey of related results. In the final section 4.5 we give a concise and streamlined proof of the result [113] illustrating the general strategy of V. S. Buslaev and G. Perelman in the case of a 1D Schrödinger equation coupled to a nonlinear oscillator.

4.1 Orthogonal Projection

This strategy arose in 1985–1992 in the pioneering work of A. Soffer and M. Weinstein [162, 163, 170]; see the review [161]. The results concern nonlinear $U(1)$-invariant Schrödinger equations with real potential $V(x)$

$$i\dot{\psi}(x,t) = -\Delta\psi(x,t) + V(x)\psi(x,t) + \lambda|\psi(x,t)|^p\psi(x,t), \qquad x \in \mathbb{R}^n,$$

(4.1.1)

where $\lambda \in \mathbb{R}$, $p = 3$ or 4, $n = 2$ or $n = 3$ and $\psi(x,t) \in \mathbb{C}$. The corresponding Hamiltonian functional reads

$$\mathcal{H} = \int \left[\frac{1}{2}|\nabla\psi|^2 + \frac{1}{2}V(x)|\psi(x)|^2 + \frac{\lambda}{p}|\psi(x)|^p \right] dx.$$

For $\lambda = 0$, the equation (4.1.1) is linear. It is assumed that the discrete spectrum of the short-range Schrödinger operator $H := -\Delta + V(x)$ is a single point $\omega_* < 0$, and the point zero is neither an eigenvalue nor a resonance for H. Let $\phi_*(x)$ denote the corresponding ground state:

$$H\phi_*(x) = \omega_*\phi_*(x).$$

(4.1.2)

Then $C\phi_*(x)e^{-i\omega_* t}$ are periodic solutions for all complex constants C. Corresponding phase curves are circles, filling the complex plane.

For nonlinear equations (4.1.1) with a small real $\lambda \neq 0$, it turns out that a wonderful *bifurcation* occurs: small neighborhood of the zero of the complex plane turns into an analytic invariant solitary manifold $\mathcal{S}$ that is still filled with invariant circles that are trajectories of *stationary orbits* of type (3.1.11),

$$\psi(x,t) = \psi_\omega(x)e^{-i\omega t}, \qquad (4.1.3)$$

whose frequencies ω are close to ω_*.

Remark 4.1.1 All these solutions $\psi_\omega(x)e^{-i\omega t}$ are called *ground states*.

The main result of [162, 163] (see also [155]) is long-time attraction to one of these ground states for any solution of equation (4.1.1) with sufficiently small $\lambda > 0$ in the case of small initial data:

$$\psi(x,t) = \psi_\pm(x)e^{-i\omega_\pm t} + r_\pm(x,t), \qquad (4.1.4)$$

where the remainder decay in weighted norms for $\sigma > 2$ is

$$\|\langle x \rangle^{-\sigma} r_\pm(\cdot,t)\|_{L^2(\mathbb{R}^n)} \to 0, \qquad t \to \pm\infty,$$

where $\langle x \rangle := (1 + |x|)^{1/2}$. The proof relies on linearization of the dynamics and decomposition of solutions into two components

$$\psi(t) = e^{-i\Theta(t)}(\psi_{\omega(t)} + \phi(t)),$$

with the orthogonality condition [162, (3.2) and (3.4)]

$$\langle \psi_{\omega(t)}, \phi(t) \rangle = 0. \qquad (4.1.5)$$

This orthogonality and dynamics (4.1.1) imply the *modulation equations* for $\omega(t)$ and $\gamma(t)$, where $\gamma(t) := \Theta(t) - \int_0^t \omega(s)ds$ (see (3.2) and (3.9a)–(3.9b) from [162]). The orthogonality (4.1.5) implies that the component $\phi(t)$ lies in the continuous spectral space of the Schrödinger operator $H(\omega_0) := -\Delta + V + \lambda|\psi_{\omega_0}|^p$, which leads to time decay of $\phi(t)$ (see [162, (4.2a) and (4.2b)]). Finally, this decay implies the convergence $\omega(t) \to \omega_\pm$ and the asymptotics (4.1.4).

These results and methods were further developed in numerous works for nonlinear Schrödinger, wave, and Klein–Gordon equations with potentials under various spectral assumptions on linearized dynamics; see [108, 113, 141, 155, 164, 165, 170].

4.2 Symplectic Projection

A genuine breakthrough in the theory of asymptotic stability was achieved in 1990–2003 by V. S. Buslaev, G. Perelman, and C. Sulem [110, 111, 112], who first extended asymptotics of type (4.1.4) to 1D translation-invariant Schrödinger equation

$$i\dot\psi(x,t) = -\psi''(x,t) - F(\psi(x,t)), \qquad x \in \mathbb{R} \tag{4.2.1}$$

without smallness conditions on the nonlinearity and initial data.

The equation is assumed to be $U(1)$-invariant, which means that the nonlinear function $F(\psi) = -\nabla_{\overline\psi}U(\psi)$ satisfies identities (3.1.8)–(3.1.10). Also, the following condition is assumed:

$$U(\psi) = \mathcal{O}(|\psi|^{10}), \qquad \psi \to 0, \tag{4.2.2}$$

which is required probably by a failure of suitable technique. Under some simple additional conditions on the potential U (see below), there exist *stationary orbits* that are finite-energy solutions of the form

$$\psi(x,t) = \psi_0(x)e^{i\omega_0 t}, \tag{4.2.3}$$

with $\omega_0 > 0$. The amplitude $\psi_0(x)$ satisfies the corresponding stationary equation

$$-\omega_0\psi_0(x) = -\psi_0''(x) - F(\psi_0(x)), \qquad x \in \mathbb{R}, \tag{4.2.4}$$

which implies the "energy conservation"

$$\frac{|\psi_0'(x)|^2}{2} + U_e(\psi_0(x)) = E, \tag{4.2.5}$$

where the "effective potential" $U_e(\psi) = U(\psi) + \omega_0\frac{|\psi|^2}{2} \sim \omega_0\frac{|\psi|^2}{2}$ as $\psi \to 0$ by (4.2.2). For the existence of a finite-energy solution (4.2.3), the graph of the effective potential $U_e(\psi)$ should be similar to Figure 4.1. The finite-energy solution is defined by (4.2.5) with the constant $E = U_e(0)$, since for other E, the solutions to (4.2.5) do not converge to zero as $|x| \to \infty$. This equation with $E = U_e(0)$ implies that

$$\frac{|\psi_0'(x)|^2}{2} = U_e(0) - U_e(\psi_0(x)) \sim \frac{\omega_0}{2}\psi_0^2(x). \tag{4.2.6}$$

Hence, for finite-energy solutions,

$$\psi_0(x) \sim e^{-\sqrt{\omega_0}|x|}, \qquad |x| \to \infty. \tag{4.2.7}$$

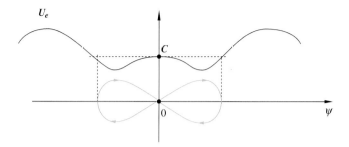

Figure 4.1 Reduced potential and soliton.

It is easy to verify that the following functions are also solutions (*moving solitons*):

$$\psi_{\omega,v,a,\theta}(x,t) = \psi_{\omega}(x - vt - a)e^{i(\omega t + kx + \theta)}, \qquad \omega = \omega_0 - v^2/4, \qquad k = v/2. \tag{4.2.8}$$

The set of all such solitons with parameters ω, v, a, θ forms a 4D smooth submanifold $\mathcal{S}$ in the Hilbert phase space $L^2(\mathbb{R})$. Moving solitons (4.2.8) are obtained from standing soliton (4.2.3) by the Galilean transformation

$$G(a,v,\theta) : \psi(x,t) \mapsto \varphi(x,t) = \psi(x - vt - a,t)e^{i\left(-\frac{v^2}{4}t + \frac{v}{2}x + \theta\right)}. \tag{4.2.9}$$

It is easy to verify that the Schrödinger equation (4.2.1) is invariant with respect to this symmetry group.

Linearization of the Schrödinger equation (4.2.1) on the stationary orbit (4.2.3) is obtained by substitution, $\psi(x,t) = (\psi_0(x) + \chi(x))e^{-i\omega_0 t}$, and retaining terms of the first order in χ. This linearized equation contains χ and $\overline{\chi}$, and hence, it is not linear over the field of complex numbers. This follows from the fact that the nonlinearity of $F(\psi)$ is not complex-analytic due to the $U(1)$-invariance (3.1.8). Complexification of this linearized equation reads

$$\dot{\Psi}(x,t) = C_0 \Psi(x,t), \tag{4.2.10}$$

where $\Psi(x,t) \in \mathbb{C}^2$.

Note that the operator $C_0 = C_{\omega_0,0,0,0}$ corresponds to the linearization on the soliton (4.2.8) which is one of the solitons (4.2.8) corresponding to parameters $\omega = \omega_0$ and $a = v = \theta = 0$. Similar operators $C_{\omega,a,v,\theta}$, corresponding to linearization at solitons (4.2.8) with various parameters ω, a, v, θ, are connected with C_0 via the differential of the Galilean transformation (4.2.9). Therefore, their spectral properties completely coincide. In particular, their continuous spectrum coincides with $(-i\infty, -i\omega_0] \cup [i\omega_0, i\infty)$.

The main results of [110] are asymptotics of type (4.1.4) for solutions with initial data close to the solitary manifold $\mathcal{S}$:

$$\psi(x,t) = \psi_\pm(x - v_\pm t)e^{-i(\omega_\pm t + k_\pm x)} + W(t)\Phi_\pm + r_\pm(x,t), \qquad \pm t > 0, \tag{4.2.11}$$

where $W(t)$ is the dynamical group of the free Schrödinger equation, $\Phi_\pm$ are some scattering states of finite-energy, and $r_\pm$ are remainder terms that decay to zero in a global norm:

$$\|r_\pm(\cdot, t)\|_{L^2(\mathbb{R})} \to 0, \qquad t \to \pm\infty. \tag{4.2.12}$$

More general asymptotics were obtained in [111, 112] under the following assumptions on the spectrum of the generator B_0:

U1 The discrete spectrum of the operator C_0 consists of exactly three eigenvalues 0 and $\pm i\lambda$, and

$$\lambda < \omega_0 < 2\lambda. \tag{4.2.13}$$

This condition means that the discrete mode can interact with the continuous spectrum already in the first order of perturbation theory.

U2 The edge points $\pm i\omega_0$ of the continuous spectrum are neither eigenvalues nor resonances of C_0.

U3 Furthermore, the condition [112, (1.0.12)] is assumed, which means a strong coupling of discrete and continuous spectral components, providing energy radiation, similarly to the Wiener condition (1.5.13). The condition [112, (1.0.12)] ensures that the interaction of a discrete component with a continuous spectrum does not vanish in the first order of perturbation theory. This condition is a nonlinear version of the Fermi Golden Rule [158], which was introduced by I. M. Sigal in the context of nonlinear PDEs [76].

In 2001, Cuccagna extended results of [110, 111, 112] to multidimensional translation-invariant Schrödinger equations in the dimensions $n \geq 2$ [114].

Method of symplectic projection in the Hilbert phase space A novel approach [110, 111, 112] relies on *symplectic projection* of solutions onto the solitary manifold. This means that

$$Z := \psi - S \quad \text{is symplectic-orthogonal to the tangent space} \quad \mathcal{T} := T_S\mathcal{S}$$

for the projection $S := P\psi$. This projection is correctly defined in a small neighborhood of $\mathcal{S}$ because $\mathcal{S}$ is a *symplectic manifold*, i.e., the corresponding symplectic form is nondegenerate on the tangent spaces $T_S\mathcal{S}$.

Thus, a solution $\psi(t)$ for each $t > 0$ decomposes as $\psi(t) = S(t) + Z(t)$, where $S(t) := P\psi(t)$, and the dynamics are linearized on the soliton $S(t)$. Similarly, for each $t \in \mathbb{R}$, the total Hilbert phase space $\mathcal{X} := L^2(\mathbb{R})$ is split as $\mathcal{X} = \mathcal{T}(t) \oplus \mathcal{Z}(t)$, where $\mathcal{Z}(t)$ is a **symplectic-orthogonal** complement to the tangent space $\mathcal{T}(t) := T_{S(t)}\mathcal{S}$. The corresponding equation for the *transversal component* $Z(t)$ reads

$$\dot{Z}(t) = A(t)Z(t) + N(t),$$

where $A(t)Z(t)$ is the linear part and $N(t) = \mathcal{O}(\|Z(t)\|^2)$ is the corresponding nonlinear part.

The main difficulties in studying this equation are as follows: (i) it is *nonautonomous* and (ii) the generators $A(t)$ *are not selfadjoint* (see the appendix in [137]). It is important that $A(t)$ are *Hamiltonian operators*, for which the existence of spectral decomposition is provided by the Krein–Langer theory of J-selfadjoint operators [145, 148]. In [137, 138] we have developed a special version of this theory providing the corresponding eigenfunction expansion, which is necessary for the justification of the approach [110, 111, 112]. The main steps of this strategy are as follows.

– **Modulation equations.** The parameters of the soliton $S(t)$
  satisfy **modulation equations**, for example, for the speed $v(t)$ we have

$$\dot{v}(t) = M(\psi(t)),$$

  where $M(\psi) = \mathcal{O}(\|Z\|^2)$ for small norms $\|Z\|$. This means that the parameters change "superslowly" near the solitary manifold, like adiabatic invariants.
– **Tangent and transversal components.** The *transversal component* $Z(t)$
  in the splitting $\psi(t) = S(t) + Z(t)$ belongs to the *transversal subspace* $\mathcal{Z}(t)$. The tangent space $\mathcal{T}(t)$ is the root space of the generator $A(t)$ and corresponds to the "unstable spectral point" $\lambda = 0$. The key observation is that

  (i) the transversal subspace $\mathcal{Z}(t)$ is **invariant** with respect to the generator $A(t)$, since the subspace $\mathcal{T}(t)$ is invariant, and $A(t)$ is the Hamiltonian operator;
  (ii) moreover, the transversal subspace $\mathcal{Z}(t)$ does not contain the tangent vectors corresponding to the unstable eigenvalue $\lambda = 0$.

– **Continuous and discrete components.** The transversal component allows further splitting $Z(t) = z(t) + f(t)$, where $z(t)$ and $f(t)$ belong, respectively, to discrete and continuous spectral subspaces $\mathcal{Z}_d(t)$ and $\mathcal{Z}_c(t)$ of $A(t)$ in the space $\mathcal{Z}(t) = \mathcal{Z}_d(t) + \mathcal{Z}_c(t)$.

– **Poincaré normal form and Fermi Golden Rule.** The component $z(t)$ satisfies a nonlinear equation, which is reduced to Poincaré normal form up to higher-order terms [112, Equations (4.3.20)]. The normal form allowed us to obtain some "conditional decay" for $z(t)$ using the Fermi Golden Rule [112, (1.0.12)]. For the relativistic Ginzburg–Landau equation, a similar reduction was done in [136, Equations (5.18)].

– **Method of majorants.** A skillful combination of the conditional decay for $z(t)$ with the superslow evolution of the soliton parameters allows one to prove the decay for $f(t)$ and $z(t)$ by the method of majorants. Finally, this decay implies the asymptotics (4.2.11)–(4.2.12).

Remark 4.2.1 (i) The role of the symplectic projection in the theory of V. S. Buslaev, G. Perelman, and C. Sulem [110, 111, 112] probably was suggested by the theory of orbital stability of M. Grillakis, J. Shatah, and W. Strauss [100, 101], which extends to Hamiltonian PDEs the stability theory of finite-dimensional Hamiltonian systems with symmetry groups; see [102, 103]. The last theory, in its own turn, dates back to H. Poincaré, who established the theory of stability of the fixed points of the reduced dynamics, which he called *relative equilibria* [104].

(ii) The difference of the theory [110, 111, 112] with [100, 101] is as follows.

 (i) The linearized dynamics in [100, 101] are stable in the transversal directions because the positive spectrum is away from zero and, hence, the conserved Hamiltonian serves as the Lyapunov function in these directions.

(ii) On the other hand, in [110, 111, 112], the positive spectrum of this transversal dynamics is not away from zero. However, the asymptotic stability holds since the positive spectrum is absolute continuous.

4.3 Generalizations and Applications

N-soliton solutions The methods and results of [112] were developed in [149, 150, 151, 152, 153, 156, 157, 159, 160] for N-soliton solutions for translation-invariant nonlinear Schrödinger equations.

Multiphoton radiation In [116] Cuccagna and Mizumachi extended methods and results of [112] to the case when the inequality (4.2.13) is changed to

$$N\lambda < \omega_0 < (N+1)\lambda,$$

with some natural $N > 1$, and the corresponding analog of condition U3 holds. It means that the interaction of discrete modes with a continuous spectrum occurs only in the Nth order of perturbation theory. The decay rate of the remainder term (4.2.12) worsens with growing N.

Linear equations coupled to nonlinear oscillators and particles The methods and results of [112] were extended (i) in [113, 141] to the Schrödinger equation coupled to a nonlinear $U(1)$-invariant oscillator, (ii) in [125, 127] to systems (2.1.1) and (1.6.1) with zero external fields, and (iii) in [126, 134, 140] to similar translation-invariant systems of the Klein–Gordon, Schrödinger, and Dirac equations coupled to a particle. The survey of these results can be found in [124].

For example, article [127] concerns solutions to the system (2.1.1) with initial data close to a solitary manifold (2.1.3) in weighted norm

$$\|\psi\|_\sigma^2 = \int \langle x \rangle^{2\sigma} |\psi(x)|^2 dx$$

with sufficiently large $\sigma > 0$. Namely, the initial state is close to soliton (2.1.3) with some parameters v_0, a_0:

$$\|\nabla\psi(x,0) - \nabla\psi_{v_0}(x-a_0)\|_\sigma + \|\psi(x,0) - \psi_{v_0}(x-a_0)\|_\sigma$$

$$+ \|\pi(x,0) - \pi_{v_0}(x-a_0)\|_\sigma + |q(0) - a_0| + |\dot{q}(0) - v_0| \le \varepsilon,$$

where $\sigma > 5$ and where $\varepsilon > 0$ is sufficiently small. Moreover, the Wiener condition (1.5.13) is assumed for $k \ne 0$. Additionally, let

$$\partial^\alpha \hat{\rho}(0) = 0, \quad |\alpha| \le 5,$$

which is equivalent to equalities

$$\int x^\alpha \rho(x)\, dx = 0, \quad |\alpha| \le 5.$$

Under these conditions, the main results of [127] are the asymptotics

$$\ddot{q}(t) \to 0, \quad \dot{q}(t) \to v_\pm, \quad q(t) \sim v_\pm t + a_\pm, \qquad t \to \pm\infty$$

(cf. (2.1.8) and (2.1.11)) and the attraction to solitons (2.1.9), where the remainder now decays in *global weighted norms* in the comoving frame (cf. (2.1.10)):

$$\|\nabla r_\pm(q(t)+x,t)\|_{-\sigma} + \|r_\pm(q(t)+x,t)\|_{-\sigma}$$

$$+ \|s_\pm(q(t)+x,t)\|_{-\sigma} \to 0, \qquad t \to \pm\infty.$$

Relativistic equations In [107, 109, 144, 135, 136] the asymptotic stability of solitary manifolds was established for the first time for *relativistic* nonlinear equations. Namely, in [107] and [144, 135, 136], asymptotics of the type (4.2.11) were obtained for 1D relativistic nonlinear wave equations (2.2.4) with potentials of the Ginzburg–Landau type and in [109] for relativistic nonlinear Dirac equations. In [139] we have constructed examples of potentials providing all spectral properties of the linearized dynamics imposed in [144, 135, 136].

In [137, 138] we have justified the eigenfunction expansions for non-self-adjoint Hamiltonian operators which were used in [144, 135, 136]. For the justification we have developed a special version of the Krein–Langer theory of J-selfadjoint operators [145, 148].

Cherenkov radiation The article [122] concerns a system of type (2.1.1) with the Schrödinger equation instead of the wave equation (system (1.9)–(1.10) in [122]). This system is considered as a model of the Cherenkov radiation. The main result of [122] is long-time convergence to a soliton with the sonic speed for initial solitons with a supersonic speed in the case of a weak interaction (the "Bogolyubov limit") and small initial field. The asymptotic stability of solitary manifolds for a very close system with the Schrödinger equation was established in [134].

4.4 Further Generalizations

The results on asymptotic stability of solitary manifolds were developed in different directions.

Systems with several bound states Articles [106, 115, 167, 168, 169] concern asymptotic stability of stationary orbits (4.1.3) for the nonlinear Schrödinger, Klein–Gordon, and wave equations in the case of several simple eigenvalues of the linearization. The typical assumptions are as follows:

(i) the endpoint of a continuous spectrum is neither an eigenvalue nor a resonance for linearized equations;
(ii) the eigenvalues of the linearized equation satisfy several nonresonance conditions;
(iii) there is an appropriate version of the Fermi Golden Rule.

One typical difficulty is the possible long stay of solutions near metastable tori which correspond to approximate resonances. Great efforts are being made to show that the role of metastable tori decreases like $t^{-1/2}$ as $t \to \infty$.

The typical result is the long-time asymptotics "ground state + dispersive wave" in the norm $H^1(\mathbb{R}^3)$ for solutions close to the ground state.

General Theory of Relativity The article [123] concerns so-called kink instability of self-similar and spherically symmetric solutions of the equations of the General Theory of Relativity with a scalar field, as well as with a "hard fluid" as sources. The authors constructed examples of self-similar solutions that are unstable to the kink perturbations.

The article [117] examines linear stability of slowly rotating Kerr solutions for the Einstein equations in a vacuum. In [201] a pointwise damping of solutions to the wave equation is investigated for the case of stationary asymptotically flat space-time in the 3D case.

In [105] the Maxwell equations are considered outside slowly rotating Kerr black hole. The main results are (i) boundedness of a positive definite energy on each hypersurface $t = \text{const}$ and (ii) convergence of each solution to a stationary Coulomb field.

In [118] the pointwise decay was proved for linear waves against the Schwarzschild black hole.

Method of concentration compactness In [130] the concentration compactness method was used for the first time to prove global well-posedness, scattering, and blowup of solutions to critical focusing nonlinear Schrödinger equation

$$i\dot{\psi}(x,t) = -\Delta\psi(x,t) - |\psi(x,t)|^{\frac{4}{n-2}}\psi(x,t), \qquad x \in \mathbb{R}^n,$$

in the radial case. Later on, these methods were extended in [119, 121, 131, 146] to general nonradial solutions and to nonlinear wave equations of the type

$$\ddot{\psi}(x,t) = \Delta\psi(x,t) + |\psi(x,t)|^{\frac{4}{n-2}}\psi(x,t), \qquad x \in \mathbb{R}^n.$$

One of the main results is splitting of the set of initial states, close to the critical energy level, into three subsets with certain long-term asymptotics: either a blowup in a finite time, or an asymptotically free wave, or the sum of the ground state and an asymptotically free wave. All three alternatives are possible; all nine combinations with $t \to \pm\infty$ are also possible. Lectures in [154] give an excellent introduction to this area. The articles [120, 132] concern supercritical nonlinear wave equations.

Recently, these methods and results were extended to critical wave mappings [129, 128, 146, 147]. The "decay onto solitons" is proved: every 1-equivariant finite-energy wave mapping of the exterior of a ball with Dirichlet boundary conditions into a 3D sphere exists globally in time and dissipates into a single stationary solution of its own topological class.

4.5 The 1D Schrödinger Equation Coupled to an Oscillator

In this section we illustrate the strategy of V. S. Buslaev and G. Perelman [110, 111] (which was also used in [112] and in many works cited in Sections 4.2, 4.3, and 4.4) by application to 1D Schrödinger equation coupled to a nonlinear oscillator, see [113]. The coupled system is invariant with respect to the phase rotation group $U(1)$. For initial states close to a stationary orbit, the solution converges to a sum of another stationary orbit and dispersive wave which is a solution to the free Schrödinger equation. The proofs are complete and rely on the strategy of [110, 111]: the linearization of the dynamics on the solitary manifold, the symplectic orthogonal projection and method of majorants.

4.5.1 Introduction

Our main goal is the study of asymptotic stability of "quantum stationary states" for a model $U(1)$-invariant nonlinear Schrödinger equation

$$i\dot{\psi}(x,t) = -\psi''(x,t) - \delta(x)F(\psi(0,t)), \quad x \in \mathbb{R}. \qquad (4.5.1)$$

Here $\psi(x,t)$ is a continuous complex-valued wave function and F is a continuous function, the dots stand for the derivatives in t and the primes in x. All derivatives and the equation are understood in the distribution sense. Physically, equation (4.5.1) describes the system of the free Schrödinger equation coupled to a nonlinear oscillator located at the point $x = 0$; F is a nonlinear "oscillator force."

We assume that $F(\psi) = -\nabla U(\psi)$ where $U(\psi) = u(|\psi|)$. Then (4.5.1) defines a $U(1)$-invariant Hamiltonian system and admits finite-energy solutions of type $\psi_\omega(x)e^{i\omega t}$ called *stationary orbits*, which correspond to *nonlinear eigenfunctions*. The stationary orbits constitute a 2D *solitary manifold* in the Hilbert phase space of finite-energy states of the equation. We prove the asymptotics of type

$$\psi(\cdot,t) \sim \psi_{\omega_\pm}e^{i\omega_\pm t} + W(t)\Phi_\pm, \quad t \to \pm\infty, \qquad (4.5.2)$$

where $W(t)$ is the dynamical group of the free Schrödinger equation, $\Phi_\pm \in C_b(\mathbb{R}) \cap L^2(\mathbb{R})$ are the corresponding asymptotic scattering states, and the remainder converges to zero as $\mathcal{O}(|t|^{-1/2})$ in the global norm of $C_b(\mathbb{R}) \cap L^2(\mathbb{R})$. Here $C_b(\mathbb{R})$ is the space of bounded continuous functions $\mathbb{R} \to \mathbb{C}$. The asymptotics hold for the solutions with initial states close to the *stable part* of the solitary manifold, extending the methods and results of [110, 111, 112] to the equation (4.5.1).

Let us note that we impose conditions which are more general than the standard ones in the following respects:

(i) We do not hypothesize any spectral properties of the linearized equation, and do not require any smallness condition on the initial state (only closeness to the solitary manifold).

(ii) The stable part of the solitary manifold is characterized by a condition on the nonlinearity (4.5.17). The relation of this to the standard criterion for orbital stability $\partial_\omega \int |\psi_\omega(x)|^2 dx > 0$ (see [100, 101] and references therein) will be discussed below.

This progress is possible on account of the simplicity of our model, which allows an exact analysis of all spectral properties of the linearization.

Let us note the following two main novelties in our approach to the uniform decay of the dynamics in transversal directions to the solitary manifold. First, we calculate exactly all needed spectral properties of corresponding generators. Second, we do not use a spectral representation of the generator. Instead, we develop the Jensen–Kato approach, applying directly the Zygmund type Lemma 6.1 (cf. [183, Lemma 10.2]) to the Laplace integral of the resolvent. We expect that the development would be promising for more general problems.

This section is organized as follows. In Section 4.5.2, some notation and definitions are given. In Section 4.5.3 we describe all nonzero stationary orbits and formulate the main theorem. The linearization on a stationary orbit is carried out in Section 4.5.4. In Sections 4.5.5 and 4.5.6, we construct the spectral representation for the linearized equation. In Section 4.5.7 we establish the time decay for the linearized equation in the continuous spectrum. In Section 4.5.9 the modulation equations for the parameters of the soliton are displayed. The decay of the transversal component is proved in Sections 4.5.10 and 4.5.11. In Section 4.5.12 we obtain the soliton asymptotics (4.5.2). In the appendix we study the resolvent of linearized equations.

In conclusion we expect that the asymptotics (4.5.2) hold for *any* finite-energy solution of the equation (4.5.1); however, this is still an open problem.

4.5.2 Notation and Definitions

We identify a complex number $\psi = \psi_1 + i\psi_2$ with the real 2D vector $(\psi_1, \psi_2) \in \mathbb{R}^2$ and assume that the vector version $\mathbf{F}$ of the oscillator force F admits a real-valued potential,

$$\mathbf{F}(\psi) = -\nabla U(\psi), \quad \psi \in \mathbb{R}^2, \quad U \in C^2(\mathbb{R}^2). \qquad (4.5.3)$$

Then (4.5.1) is formally a Hamiltonian system with Hamiltonian

$$\mathcal{H}(\psi) = \frac{1}{2} \int |\psi'|^2 dx + U(\psi(0)), \qquad (4.5.4)$$

which is conserved for sufficiently regular finite-energy solutions. We assume that the potential $U(\psi)$ satisfies the inequality

$$U(z) \geq A - B|z|^2 \quad \text{with some} \quad A \in \mathbb{R}, \quad B > 0. \qquad (4.5.5)$$

Our key assumption concerns the $U(1)$-invariance of the oscillator, where $U(1)$ stands for the rotation group $e^{i\theta}$, $\theta \in [0, 2\pi]$ acting by phase rotation $\psi \mapsto e^{i\theta} \psi$. Namely, we assume that

$$U(\psi) = u(|\psi|^2), \qquad u \in C^2(\mathbb{R}) \qquad (4.5.6)$$

(cf. [24, 25]). In this case,

$$F(\psi) = a(|\psi|^2)\psi, \qquad \psi \in \mathbb{C}, \qquad a \in C^1(\mathbb{R}). \qquad (4.5.7)$$

Therefore,

$$F(e^{i\theta}\psi) = e^{i\theta} F(\psi), \quad \theta \in [0, 2\pi], \qquad (4.5.8)$$

and $F(0) = 0$. This rotation symmetry implies that $e^{i\theta}\psi(x,t)$ is a solution to (4.5.1) if $\psi(x,t)$ is. The equation is $U(1)$-invariant in the sense of [100, 101], and the Nöther theorem implies the *charge conservation*:

$$Q(\psi) = \int |\psi|^2 dx = \text{const.} \qquad (4.5.9)$$

The main subject of this section is an analysis of asymptotic stability of "quantum stationary orbits," or *solitary waves* in the sense of [100, 101], which are finite-energy solutions of the form

$$\psi(x,t) = \psi_\omega(x)e^{i\omega t}, \quad \omega \in \mathbb{R}. \qquad (4.5.10)$$

The frequency ω and the amplitude $\psi_\omega(x)$ solve the following *nonlinear eigenvalue problem*:

$$-\omega\psi_\omega(x) = -\psi_\omega''(x) - \delta(x)F(\psi_\omega(0)), \quad x \in \mathbb{R}, \qquad (4.5.11)$$

which follows directly from (4.5.1) and (4.5.7), since $\omega \in \mathbb{R}$.

Definition 4.5.1 $\mathcal{S}$ denotes the set of all nonzero solutions $\psi_\omega(x) \in H^1(\mathbb{R})$ to (4.5.11) with all possible $\omega \in \mathbb{R}$.

Here $H^1(\mathbb{R}) = H^1$ denotes the Sobolev space of complex valued measurable functions with $\int (|\psi'|^2 + |\psi|^2) dx < \infty$. We give below in Section 4.5.3 a complete analysis of the set S of all nonzero stationary orbits $\psi_\omega(x)$ by an explicit calculation: it consists of functions $C(\omega) e^{-\sqrt{\omega}|x| + i\theta}$ with $C > 0$, $\omega = \omega(C) > 0$ and any $\theta \in [0, 2\pi]$, and C restricted to lie in a set which, in the case of polynomial F, is a finite union of 1D intervals. Notice that $C = 0$ corresponds to the zero function $\psi(x) = 0$, which is always a solitary wave as $F(0) = 0$, and for $\omega \le 0$, only the zero stationary orbit exists.

Our main results describe the large time behavior of the global solutions whose existence is guaranteed by the following theorem, which is proved in [133].

Theorem 4.5.2 *(i) Let conditions (4.5.3) and (4.5.5) hold. Then, for any initial state $\psi(0) \in H^1$, there exists a unique solution $\psi(\cdot) \in C_b(\mathbb{R}, H^1)$ to the equation (4.5.1).*
(ii) The following a priori bound holds:

$$\sup_{t \in \mathbb{R}} \|\psi(t)\|_{H^1} < \infty. \tag{4.5.12}$$

The functional spaces we are going to consider are the weighted Banach spaces L^p_β, $p \in [1, \infty)$, $\beta \in \mathbb{R}$ of complex-valued measurable functions with the norm

$$\|u\|_{L^p_\beta} = \|(1 + |x|)^\beta u(x)\|_{L^p}. \tag{4.5.13}$$

4.5.3 Stationary Orbits and the Main Theorem

Lemma 4.5.3 *The set of all nonzero stationary orbits is given by*

$$S = \Big\{ \psi_\omega e^{i\theta} = C e^{i\theta - \sqrt{\omega}|x|} : \omega > 0, \quad C > 0,$$
$$\sqrt{\omega} = a(C^2)/2 > 0, \quad \theta \in [0, 2\pi] \Big\}.$$

Proof Let us calculate all stationary orbits (4.5.10). The equation (4.5.11) implies $\psi''(x) = \omega \psi(x)$, $x \ne 0$, hence the formula $\psi(x) = C_\pm e^{\sqrt{\omega}x}$ gives two linearly independent solutions in each of the two regions $\pm x > 0$ depending on which branch of $\sqrt{\omega}$ is chosen. Since $\psi(x) \in L^2$ it is necessary that $\omega > 0$ and the branch is chosen with $\pm\sqrt{\omega} > 0$ for $\pm x < 0$. Furthermore, since $\psi'(x) \in L^2$, the function $\psi(x)$ is continuous, hence $C_- = C_+ = C$ and the solutions are of the form

$$\psi(x) = C e^{-\varkappa|x|}, \quad \varkappa = \sqrt{\omega} > 0, \quad \omega > 0. \tag{4.5.14}$$

Finally we get an algebraic equation for the constant C equating the coefficients of $\delta(x)$ in both sides of (4.5.11):

$$0 = \psi'(0+) - \psi'(0-) + F(\psi(0)). \tag{4.5.15}$$

This implies $0 = -2\varkappa C + F(C)$, or equivalently,

$$\varkappa = \frac{F(C)}{2C} = \frac{a(C^2)}{2}. \tag{4.5.16}$$

$\square$

Corollary 4.5.4 *The set $\mathcal{S}$ is a smooth manifold with co-ordinates $\theta \in \mathbb{R}$ mod 2π and $C > 0$ such that $a(C^2) > 0$.*

Remark 4.5.5 We will analyze only the stationary orbits with $a'(C) \neq 0$. On the manifold $\mathcal{S}$ we have $\omega = \varkappa^2$ with $\varkappa = a(C^2)/2$ according to (4.5.16). Hence, the parameters θ, ω locally also are smooth coordinates on $\mathcal{S}$ at the points with $a' = a'(C) \neq 0$ since $\omega' = 2\varkappa\varkappa' = aa'C \neq 0$ then; see Figure 2.

The stationary orbits are a trajectory $\psi_{\omega(t)}(x)e^{i\theta(t)} = Ce^{-\sqrt{\omega(t)}|x|}e^{i\theta(t)}$, where the parameters satisfy the equation $\dot\theta = \omega$, $\dot\omega = 0$. The stationary orbit $t \mapsto e^{i\omega t}\psi_\omega(x)$ maps out in time an orbit $\theta \mapsto e^{i\theta}\psi_\omega(x)$ of the $U(1)$ symmetry group. This group acts on the Hilbert phase space $H^1(\mathbf{R})$ preserving the Hamiltonian (4.5.4) and the symplectic form (4.5.40); in other words the stationary orbits (4.5.10) are relative equilibria of the corresponding Hamiltonian system [104].

Let us denote $N(C) = \mathcal{Q}(\psi_\omega(x))$ with $\omega = \varkappa^2$, and $\varkappa = a(C^2)/2$ according to (4.5.16). It is easy to compute that $N(C) = C^2/\varkappa$. We now differentiate:

$$N'(C) = \frac{2C}{\varkappa} - \frac{C^2\varkappa'}{\varkappa^2}.$$

Differentiating the identity (4.5.16), we obtain $\varkappa' = a'C$. Thus, again by (4.5.16),

$$N'(C) = \frac{2C}{\varkappa}\left(1 - \frac{a'C^2}{a}\right) \neq 0$$

if $C > 0$, $a > 0$ and $a' \neq a/C^2$. Therefore noticing that $N'(C) = \omega'(C)$ $\partial_\omega \mathcal{Q}(\psi_\omega)$ with $\omega'(C) = 2\varkappa\varkappa' = aa'C$, we obtain the following result.

Lemma 4.5.6 *For $C > 0$, $a > 0$, we have*

$$\partial_\omega \mathcal{Q}(\psi_\omega) < 0 \quad \text{if } a' \in (-\infty, 0) \cup (a/C^2, +\infty),$$

and

$$\partial_\omega \mathcal{Q}(\psi_\omega) > 0 \quad \text{if } 0 < a' < a/C^2.$$

Remark 4.5.7 (i) Orbital stability of stationary orbits is a much studied subject (see [100, 101] for very general theorems in this area, and [166] for an approach more similar to that taken in this section). The standard condition for orbital stability ([100, 101]) for the present problem would read $\partial_\omega \mathcal{Q}(\psi_\omega) > 0$; this is expected to be a necessary and sufficient condition for orbital stability when the Hessian of the augmented Hamiltonian ([166]) has a single negative eigenvalue. In the present problem it can be easily calculated that this Hessian is non-negative when $a' < 0$ and thus the standard condition is not necessarily relevant if $a' < 0$. Indeed Theorem 4.5.9 asserts stability in the case $a' < 0$. Restricting to $a' > 0$, in which case the Hessian does have a single negative eigenvalue, the calculation above shows that orbital stability is expected to hold when $a' < a/C^2$. In this section we will work under the spectral condition (4.5.18) which, for $a' > 0$, is slightly stricter: it is imposed to ensure that the linearization has no discrete spectrum except zero (which is always present on account of the circular symmetry of the problem). If $a/\sqrt{2}C^2 < a' < a/C^2$ there are two purely imaginary eigenvalues of the linearized operator. It is intended to treat this case in a later publication thus extending our proof of asymptotic stability to the entire range

$$-\infty < a' < a/C^2. \tag{4.5.17}$$

For $a' > a/C^2$ the linearized operator has a positive eigenvalue and the stationary orbit is linearly unstable.

(ii) It is explained at the end of Section 4.5.4 that (4.5.6) can be interpreted as saying the restriction of the symplectic form (4.5.40) to the tangent space to $\mathcal{S}$ is nondegenerate (i.e., $\mathcal{S}$ satisfies the condition to be a symplectic submanifold).

Definition 4.5.8 We say the stationary orbit $\psi_\omega(x)e^{i\theta} = Ce^{-\sqrt{\omega}|x|+i\theta}$, $C > 0$ satisfies the spectral condition if $\omega > 0$ and (cf. Remark 4.5.5)

$$a'(C^2) \in (-\infty, 0) \cup (0, a(C^2)/(\sqrt{2}C^2)). \tag{4.5.18}$$

Let us denote by $W(t)$ the dynamical group of the free Schrödinger equation: $W(t)f$ is defined by the Fourier representation for all tempered distributions f. Our main theorem is the following:

Theorem 4.5.9 *Let conditions (4.5.3), (4.5.5), and (4.5.6) hold, $\beta \geq 2$ and $\psi(x,t) \in C(\mathbb{R}, H^1)$ be the solution to the equation (4.5.1) with initial state*

$\psi(0) \in H^1 \cap L_\beta^1$, which is close to a stationary orbit $\psi_{\omega_0} e^{i\theta_0} = C_0 e^{-\sqrt{\omega_0}|x|+i\theta_0}$ with $C_0 > 0$ and $\omega_0 > 0$:

$$d := \|\psi(0) - \psi_{\omega_0} e^{i\theta_0}\|_{H^1 \cap L_\beta^1} \ll 1. \qquad (4.5.19)$$

Assume further that the spectral condition (4.5.18) holds for the stationary orbit with $C = C_0$. Then for sufficiently small $d > 0$ the solution admits the following asymptotics:

$$\psi(\cdot, t) = \psi_{\omega_\pm} e^{i\omega_\pm t} + W(t)\Phi_\pm + r_\pm(t), \quad t \to \pm\infty, \qquad (4.5.20)$$

where $\Phi_\pm \in C_b(\mathbb{R}) \cap L^2(\mathbb{R})$ are the corresponding asymptotic scattering states, and

$$\|r_\pm(t)\|_{C_b(\mathbb{R}) \cap L^2(\mathbb{R})} = \mathcal{O}(|t|^{-1/2}), \quad t \to \pm\infty. \qquad (4.5.21)$$

Remark 4.5.10 It is possible to derive further information about the structure of $\Phi_\pm$ and $r_\pm(t)$, as discussed towards the end of Section 4.5.10.

4.5.4 Linearization on the Stationary Orbit

As the first step in the proof of the main theorem, let us linearize the nonlinear Schrödinger equation (4.5.1) on a stationary orbit $e^{i(\omega t + \theta)}\psi_\omega(x)$, with $\psi_\omega(x) = Ce^{-\varkappa|x|}$ where $\varkappa = \sqrt{\omega} > 0$ and $C > 0$. Substituting

$$\psi(x, t) = e^{i(\omega t + \theta)}(\psi_\omega(x) + \chi(x, t)) \qquad (4.5.22)$$

to (4.5.1), we obtain

$$-\omega\chi(x, t) + i\dot\chi(x, t) = -\chi''(x, t) - \delta(x)[F(C + \chi(0, t)) - F(C)]. \qquad (4.5.23)$$

Use the representation (4.5.7) to write

$$
\begin{aligned}
F(C + \chi) - F(C) &= a(|C + \chi|^2)(C + \chi) - a(|C|^2)C \\
&= a((C + \chi)(\overline{C} + \overline{\chi}))(C + \chi) - a(|C|^2)C \\
&= a(|C|^2)\chi + a'(|C|^2)C(C\overline{\chi} + \overline{C}\chi) + \mathcal{O}(|\chi|^2) \\
&= a(C^2)\chi + a'(C^2)C^2(\overline{\chi} + \chi) + \mathcal{O}(|\chi|^2) \qquad (4.5.24)
\end{aligned}
$$

since $C \geq 0$. Hence, the first order part of (4.5.23) is given by

$$i\dot\chi(x, t) = -\chi''(x, t) + \omega\chi(x, t)$$

$$- \delta(x)[a(C^2)\chi(0, t) + a'(C^2)C^2 2\text{Re}\,\chi(0, t)]. \qquad (4.5.25)$$

Now it is evident that the first order part is not linear over the complex field. On the other hand, it is linear over the real field. Hence, it would be useful to rewrite (4.5.25) in the real form. Namely, identify $\chi = \chi_1 + i\chi_2 \in \mathbb{C}$ with the real vector $(\chi_1, \chi_2) \in \mathbb{R}^2$ and denote it again by χ. Then (4.5.25) becomes the system

$$j\dot{\chi}(x,t) = -\chi''(x,t) + \omega\chi(x,t)$$

$$- \delta(x)[a(C^2)E + 2a'(C^2)C^2 P_1]\chi(0,t), \qquad (4.5.26)$$

where E is the unit 2×2 matrix, P_1 is the projector in $\mathbb{R}^2$ acting as $\left(\begin{smallmatrix}\chi_1\\\chi_2\end{smallmatrix}\right) \mapsto \left(\begin{smallmatrix}\chi_1\\0\end{smallmatrix}\right)$ and j is the 2×2 matrix

$$j = \begin{pmatrix} 0 & -1 \\ 1 & 0 \end{pmatrix}. \qquad (4.5.27)$$

Respectively, we also rewrite (4.5.1) in the real form

$$j\dot{\psi}(x,t) = -\psi''(x,t) - \delta(x)\mathbf{F}(\psi(0,t)), \qquad (4.5.28)$$

as an equation for $\mathbb{R}^2$-valued function $\psi(x,t)$ with $\mathbf{F}(\psi) \in \mathbb{R}^2$ which is the real vector version of $F(\psi) \in \mathbb{C}$. Then the linearization (4.5.26) reads as the system

$$j\dot{\chi}(x,t) = -\chi''(x,t) + \omega\chi(x,t) - \delta(x)\mathbf{F}'((C,0))\chi(0,t), \qquad (4.5.29)$$

where $\mathbf{F}'$ is the differential of the map $\mathbf{F}: \mathbb{R}^2 \to \mathbb{R}^2$,

$$\mathbf{F}'((C,0)) = aE + bP_1, \qquad a := a(C^2), \quad b := 2a'(C^2)C^2. \qquad (4.5.30)$$

In order to apply the Laplace transform the next step is to complexify the system (4.5.29), i.e., to consider it as a system of equations for the complex functions $\chi_1(x,t), \chi_2(x,t)$, so $\chi(x,t) \in \mathbb{C}^2$ for any fixed (x,t). This gives a system which is linear over the complex field allowing application of the Laplace transform. To write this system more concisely, let us denote the linear operator

$$\mathbf{B} := -\frac{d^2}{dx^2} + \omega - \delta(x)\mathbf{F}'((C,0)) = \begin{pmatrix} \mathbf{D}_1 & 0 \\ 0 & \mathbf{D}_2 \end{pmatrix},$$

where

$$\mathbf{D}_1 = -\frac{d^2}{dx^2} + \omega - \delta(x)[a+b],$$

$$\mathbf{D}_2 = -\frac{d^2}{dx^2} + \omega - \delta(x)a. \qquad (4.5.31)$$

The system (4.5.29) then reads as

$$\dot{\chi}(x,t) = \mathbf{C}\chi(x,t), \quad \mathbf{C} := j^{-1}\mathbf{B} = \begin{pmatrix} 0 & \mathbf{D}_2 \\ -\mathbf{D}_1 & 0 \end{pmatrix}. \tag{4.5.32}$$

Theorem 4.5.2 generalizes to the equation (4.5.32): the equation admits unique solution $\chi(x,t) \in C_b(\mathbb{R}, H^1)$ for every initial function $\chi(x,0) = \chi_0 \in H^1$. Denote by $e^{\mathbf{C}t}$ the dynamical group of equation (4.5.32) acting in the space H^1.

4.5.5 Laplace Transform

Equation (4.5.32) can be solved by the Laplace transform $\tilde{\chi}(x,\omega) := \int_0^\infty e^{-\lambda t}\chi(x,t)dt$. The Laplace transform is analytic function in the complex halfplane Re $\lambda > 0$ with the values in H^1 since the solution is bounded in H^1. This implies that the resolvent $\mathbf{R}(\lambda) := (\mathbf{C} - \lambda)^{-1}$ is also analytic for Re $\lambda > 0$, with values in the space of bounded operators on H^1. From the inversion of the Laplace transform we obtain

$$e^{\mathbf{C}t} = -\frac{1}{2\pi i}\int_{-i\infty}^{i\infty} e^{\lambda t}\mathbf{R}(\lambda + \varepsilon)\,d\lambda, \qquad t > 0, \tag{4.5.33}$$

for any $\varepsilon > 0$, where the integral converges in the sense of distributions of $t \in \mathbb{R}$.

We assume that the spectral condition (4.5.18) holds from now on. Then the resolvent admits analytic continuation from Re $\lambda > 0$ to the complex plane with the cuts $\mathcal{C}_+ = [i\omega, i\infty)$, $\mathcal{C}_- = (-i\infty, -i\omega]$, and with the pole of order two at $\lambda = 0$ as detailed in Section 4.5.13. Furthermore, for $\lambda \in \mathcal{C}_+ \cup \mathcal{C}_-$, the resolvent $\mathbf{R}(\lambda \pm \varepsilon)$ has right and left limits $\mathbf{R}(\lambda \pm 0)$ as $\varepsilon \to 0$. Then (4.5.33) implies that for any $r \in (0, \omega)$

$$e^{\mathbf{C}t} = -\frac{1}{2\pi i}\int_{|\lambda|=r} e^{\lambda t}\mathbf{R}(\lambda)\,d\lambda - \frac{1}{2\pi i}\int_{\mathcal{C}_+\cup\mathcal{C}_-} e^{\lambda t}\left(\mathbf{R}(\lambda + 0) - \mathbf{R}(\lambda - 0)\right)\,d\lambda$$

$$\tag{4.5.34}$$

by the Cauchy theorem. Setting $t = 0$, we obtain that

$$1 = -\frac{1}{2\pi i}\int_{|\lambda|=r} \mathbf{R}(\lambda)\,d\lambda - \frac{1}{2\pi i}\int_{\mathcal{C}_+\cup\mathcal{C}_-}\left(\mathbf{R}(\lambda + 0) - \mathbf{R}(\lambda - 0)\right)\,d\lambda = \mathbf{P}^0 + \mathbf{P}^c,$$

$$\tag{4.5.35}$$

where $\mathbf{P}^0$ and $\mathbf{P}^c$ stands for the corresponding Riesz projections (see [14]) onto, respectively, the generalized null space of $\mathbf{C}$, and onto the continuous spectral subspace. We will show in the next section that $\mathbf{P}^0$ is the symplectic projection, and therefore, $\mathbf{P}^c$ is also the symplectic projection. The projectors $\mathbf{P}^0, \mathbf{P}^c$ commute with $\mathbf{C}$ and with the group $e^{\mathbf{C}t}$. Let us note that

$$
\begin{cases}
\mathbf{P}^0 e^{\mathbf{C}t} = -\frac{1}{2\pi i} \int_{|\lambda|=r} e^{\lambda t} \mathbf{R}(\lambda) \, d\lambda, \\
\mathbf{P}^c e^{\mathbf{C}t} = -\frac{1}{2\pi i} \int_{\mathcal{C}_+ \cup \mathcal{C}_-} e^{\lambda t} \big(\mathbf{R}(\lambda+0) - \mathbf{R}(\lambda-0)\big) \, d\lambda
\end{cases} .
\tag{4.5.36}
$$

The first equation holds since both sides are one-parameter groups of operators, and their derivatives at $t=0$ coincide. The second equation follows from (4.5.34) and the fact that $1 = \mathbf{P}^0 + \mathbf{P}^c$ by (4.5.35). Therefore, (4.5.34) becomes

$$
e^{\mathbf{C}t} = \mathbf{P}^0 e^{\mathbf{C}t} + \mathbf{P}^c e^{\mathbf{C}t}.
\tag{4.5.37}
$$

4.5.6 Invariant Subspace of Discrete Spectrum

Here we prove that $\mathbf{P}^0$ is the symplectic projection onto the tangent space of the solitary manifold $\mathcal{S}$ at the stationary orbit $e^{j\theta}\psi_\omega$. The real form of the stationary orbit is $e^{j\theta}\Phi_\omega$ where $\Phi_\omega = (\psi_\omega(x), 0)$. The tangent space to $\mathcal{S}$ at the point $e^{j\theta}\Phi_\omega$ with parameters ω, θ is the linear span of the derivatives with respect to θ and ω (cf. Remark 4.5.5), i.e.,

$$
T_{\omega,\theta}\mathcal{S} \equiv \text{linear span}\Big\{ je^{j\theta}\Phi_\omega(x), e^{j\theta}\partial_\omega\Phi_\omega(x) \Big\}.
$$

Notice that the operator $\mathbf{C}$ corresponds to $\theta = 0$ since we have extracted the phase factors $e^{i\theta}$ from the solution in the process of linearization (4.5.22). The tangent space to $\mathcal{S}$ at the point Φ_ω with parameters $(\omega, 0)$ is spanned by the vectors

$$
T_0(\omega) := j\Phi_\omega, \quad T_1(\omega) := \partial_\omega\Phi_\omega.
\tag{4.5.38}
$$

Observe that (4.5.11) and its derivative in ω give the following identities:

$$
\mathbf{D}_2\psi_\omega = 0, \qquad \mathbf{D}_1(\partial_\omega\psi_\omega) = -\psi_\omega.
\tag{4.5.39}
$$

These formulas imply that the vectors T_0 and T_1 lie in the generalized null space of the non-selfadjoint operator $\mathbf{C}$ defined in (4.5.32) and in fact Theorem 4.5.38(ii) implies:

Lemma 4.5.11 *Let the spectral condition (4.5.18) hold. Then the generalized null space of $\mathbf{C}$ is two dimensional, is spanned by T_0, T_1, and*

$$
\mathbf{C}T_0 = 0, \qquad \mathbf{C}T_1 = T_0.
$$

We also introduce the symplectic form Ω for the real vectors ψ and η by the integral

$$\Omega(\psi, \eta) = \int \langle j\psi, \eta \rangle dx = \int (\psi_1 \eta_2 - \psi_2 \eta_1) dx, \qquad (4.5.40)$$

where $\langle \cdot, \cdot \rangle$ stands for the scalar product in $\mathbb{R}^2$. Since $a' \neq a/C^2$ then by Lemma 4.5.6

$$\mu_\omega = -\Omega(T_0, T_1) = \frac{1}{2} \partial_\omega \int |\psi_\omega|^2 dx \neq 0. \qquad (4.5.41)$$

Hence, the symplectic form Ω is nondegenerate on the tangent space $T_{\omega,0}\mathcal{S}$, i.e., $T_{\omega,0}\mathcal{S}$ is a symplectic subspace. Therefore, there exists a symplectic projection operator from $L^2(\mathbb{R})$ onto $T_{\omega,0}\mathcal{S}$.

Lemma 4.5.12 *The operator $\mathbf{P}^0$, defined in (4.5.35), is precisely the symplectic projection from $L^2(\mathbb{R})$ onto $T_{\omega,0}\mathcal{S}$, and, furthermore, it may be represented by the formula*

$$\mathbf{P}^0 \psi = b_0 T_0 + b_1 T_1 \quad \text{with} \quad -\mu_\omega b_0 = \Omega(\psi, T_1), \quad \mu_\omega b_1 = \Omega(\psi, T_0). \qquad (4.5.42)$$

Proof The coincidence of both definition (4.5.35) and (4.5.42) of operator $\mathbf{P}^0$ follows by the Cauchy residue theorem from the formulas (4.5.45)–(4.5.47) for the resolvent. $\qquad \square$

Corollary 4.5.13 $\mathbf{P}^c = 1 - \mathbf{P}^0$ *is also symplectic projection.*

Remark 4.5.14 Since $T_0(\omega), T_1(\omega)$ lie in $H^1(\mathbb{R})$ the operator $\mathbf{P}^0$ extends uniquely to define a continuous linear map $H^{-1}(\mathbb{R}) \to T_{\omega,0}\mathcal{S}$, which is still designated $\mathbf{P}^0$. In particular this operator can be applied to the Dirac measure $\delta(x)$.

Using the Taylor expansion for the $e^{\lambda t}$ at $\lambda = 0$ and the identity $\lambda \mathbf{R}(\lambda) = C\mathbf{R}(\lambda) - 1$, we obtain by (4.5.36)

$$\mathbf{P}^0 e^{Ct} = (1 + Ct)\mathbf{P}^0. \qquad (4.5.43)$$

Remark 4.5.15 On the generalized null space itself $C^2 = 0$ by Lemma 4.5.11 and so the group e^{tC} reduces to $1 + Ct$ as usual for the exponential of the nilpotent part of an operator.

4.5.7 Time Decay in Continuous Spectrum

From formulas (4.5.37) and (4.5.43) we see that the solutions $\chi(t) = e^{Ct}\chi_0$ of the linearized equation (4.5.32) do not decay as $t \to \infty$ if $\mathbf{P}^0 \chi_0 \neq 0$.

On the other hand, we do expect time decay of $\mathbf{P}^c \chi(t)$, as a consequence of the Laplace representation (4.5.36) for $\mathbf{P}^c e^{\mathbf{C}t}$:

$$\mathbf{P}^c e^{\mathbf{C}t} = -\frac{1}{2\pi i} \int_{\mathcal{C}_+ \cup \mathcal{C}_-} e^{\lambda t} \left(\mathbf{R}(\lambda + 0) - \mathbf{R}(\lambda - 0) \right) d\lambda. \qquad (4.5.44)$$

The decay for the oscillatory integral is obtained from the analytic properties of $\mathbf{R}(\lambda)$ for $\lambda \in \mathcal{C}_+ \cup \mathcal{C}_-$. The resolvent $\mathbf{R}(\lambda)$ is an integral operator with matrix-valued integral kernel

$$\mathbf{R}(\lambda, x, y) = \Gamma(\lambda, x, y) + P(\lambda, x, y), \qquad (4.5.45)$$

where the columns of matrices Γ and P are given in (4.5.134), (4.5.135), (4.5.137), (4.5.138):

$$\Gamma(\lambda, x, y) = \begin{pmatrix} \dfrac{1}{4k_+} & -\dfrac{1}{4k_-} \\ \dfrac{i}{4k_+} & \dfrac{i}{4k_-} \end{pmatrix}$$

$$\times \begin{pmatrix} e^{ik_+|x-y|} - e^{ik_+(|x|+|y|)} & -i(e^{ik_+|x-y|} - e^{ik_+(|x|+|y|)}) \\ e^{ik_-|x-y|} - e^{ik_-(|x|+|y|)} & i(e^{ik_-|x-y|} - e^{ik_-(|x|+|y|)}) \end{pmatrix},$$

$$(4.5.46)$$

$$P(\lambda, x, y) = \frac{1}{2D} \begin{pmatrix} e^{ik_+|x|} & e^{ik_-|x|} \\ i e^{ik_+|x|} & -i e^{ik_-|x|} \end{pmatrix} \begin{pmatrix} i\alpha - 2k_- & i\beta \\ -i\beta & -i\alpha + 2k_+ \end{pmatrix}$$

$$\times \begin{pmatrix} e^{ik_+|y|} & -i e^{ik_+|y|} \\ e^{ik_-|y|} & i e^{ik_-|y|} \end{pmatrix}. \qquad (4.5.47)$$

Here $k_\pm(\lambda) = \sqrt{-\omega \mp i\lambda}$ is the square root defined with cuts in the complex λ plane so that $k_\pm(\lambda)$ is an analytic function on $\mathbb{C} \setminus \mathcal{C}_\pm$ and $\operatorname{Im} k_\pm(\lambda) > 0$ for $\lambda \in \mathbb{C} \setminus \mathcal{C}_\pm$. The constants α, β, and $D = D(\lambda)$ are given by the formulas

$$\alpha = a + b/2, \ \beta = b/2, \ D = 2i\alpha(k_+ + k_-) - 4k_+ k_- + \alpha^2 - \beta^2.$$

Recall from Section 4.5.13 that $D(\lambda) \neq 0$ for $\lambda \in \mathcal{C}_+ \cup \mathcal{C}_-$. Clearly in order to understand the decay of $\mathbf{P}^c e^{t\mathbf{C}}$, it is crucial to study the behavior of $\mathbf{R}(\lambda, x, y)$ near the branch points $\lambda = \pm i\omega$ (where $k_\pm$ vanish).

We deduce time decay for the group $\mathbf{P}^c e^{t\mathbf{C}}$ by means of the following version of Lemma 10.2 from [183], which is itself based on Zygmund's lemma [191, p.45].

Let $\mathcal{F} \colon [0, \infty) \to \mathbf{B}$ be a C^2 function with values in a Banach space $\mathbf{B}$. Let us define the $\mathbf{B}$-valued function

$$I(t) = \int_0^\infty e^{-it\zeta} \mathcal{F}(\zeta)\, d\zeta.$$

Lemma 4.5.16 *Suppose that $\mathcal{F}(0) = 0$, and for some $\delta > 0$,*

$$\mathcal{F}'' \in L^1(\delta, \infty; \mathbf{B}), \tag{4.5.48}$$

and

$$\mathcal{F}''(\zeta) = \mathcal{O}(\zeta^{p-2}), \quad \zeta \downarrow 0 \tag{4.5.49}$$

in the norm of $\mathbf{B}$ for some $p \in (0, 1)$. Then $I(t) \in C_b(\varepsilon, \infty; \mathbf{B})$ for any $\varepsilon > 0$, and

$$I(t) = \mathcal{O}(t^{-1-p}) \quad \text{as} \quad t \to \infty$$

in the norm of $\mathbf{B}$.

For $\beta \geq 2$ let us introduce a Banach space $\mathcal{M}_\beta$, which is the subset of distributions which are linear combinations of L^1_β functions and multiples of the Dirac distribution at the origin with the norm:

$$\|\psi + C\delta(x)\|_{\mathcal{M}_\beta} := \|\psi\|_{L^1_\beta} + |C|. \tag{4.5.50}$$

We will apply Lemma 4.5.16 to the function $\mathcal{F}(\lambda) = \mathbf{R}(\lambda + 0) - \mathbf{R}(\lambda - 0)$ with values in the Banach space $\mathcal{B} = B(\mathcal{M}_\beta, L^\infty_{-\beta})$, the space of continuous linear maps $\mathcal{M}_\beta \to L^\infty_{-\beta}$ for any $\beta \geq 2$.

Theorem 4.5.17 *Assume that the spectral condition (4.5.18) holds so that $\lambda = 0$ is the only point in the discrete spectrum of the operator $\mathbf{C} = \mathbf{C}(\omega)$. Then for $\beta \geq 2$,*

$$\|\mathbf{P}^c e^{\mathbf{C}t}\|_{\mathcal{B}} = \mathcal{O}(t^{-3/2}), \quad t \to \infty. \tag{4.5.51}$$

First, we use the formulas (4.5.44) and (4.5.45) to obtain

$$-2\pi i \mathbf{P}^c e^{\mathbf{C}t} = \int_{\mathcal{C}_+ \cup \mathcal{C}_-} e^{\lambda t} (\Gamma(\lambda + 0) - \Gamma(\lambda - 0))\, d\lambda$$

$$+ \int_{\mathcal{C}_+ \cup \mathcal{C}_-} e^{\lambda t} (P(\lambda + 0) - P(\lambda - 0))\, d\lambda. \tag{4.5.52}$$

Next we apply Lemma 4.5.16 to each summand in the RHS of (4.5.52) separately. Then Theorem 4.5.17 immediately follows from the two lemmas below.

Lemma 4.5.18 *If the assumption of Theorem 4.5.17 holds then*

$$\int_{\mathcal{C}_+\cup\mathcal{C}_-} e^{\lambda t}(\Gamma(\lambda+0)-\Gamma(\lambda-0))\,d\lambda = \mathcal{O}(t^{-3/2}), \quad t\to\infty \qquad (4.5.53)$$

in the norm $\mathcal{B}$.

Proof We consider only the integral over $\mathcal{C}_+$ since the integral over $\mathcal{C}_-$ can be handled in the same way. The point $\lambda = i\omega$ is the branch point for k_+, therefore, if $\lambda \in \mathcal{C}_+$ then since k_- is continuous across $\mathcal{C}_+$

$$\Gamma(\lambda+0)-\Gamma(\lambda-0)=\Gamma^+(\lambda+0)-\Gamma^+(\lambda-0),$$

where Γ^+ is the sum of those terms in Γ which involve k_+. Let us consider, for example, Γ^+_{11}. The expression (4.5.46) implies for $y > 0$ that

$$\Gamma^+_{11}(\lambda,x,y) = \begin{cases} 0, & x \le 0, \\[2mm] \dfrac{e^{ik_+y}(e^{-ik_+x}-e^{ik_+x})}{4k_+}, & 0 \le x \le y, \\[3mm] \dfrac{e^{ik_+x}(e^{-ik_+y}-e^{ik_+y})}{4k_+}, & x \ge y. \end{cases}$$

For $\lambda \in \mathcal{C}_+$, the root $k_+ = \sqrt{-\omega-i\lambda}$ is real, and $k_+(\lambda+0)=-k_+(\lambda-0)$. Then, for $y > 0$,

$$\Gamma^+_{11}(\lambda+0,x,y)-\Gamma^+_{11}(\lambda-0,x,y) = -\Theta(x)\frac{\sin(\sqrt{\zeta}|x|)\sin(\sqrt{\zeta}|y|)}{\sqrt{\zeta}},$$

$$(4.5.54)$$

where $\zeta = -\omega - i\lambda$, and $\Theta(x) = 1$ for $x > 0$ and zero otherwise. The second derivative of the function $f(\zeta) = \dfrac{\sin(\sqrt{\zeta}|x|)\sin(\sqrt{\zeta}|y|)}{\sqrt{\zeta}}$ satisfies

$$|f''(\zeta)| = \left| -\frac{\sin(\sqrt{\zeta}|x|)\sin(\sqrt{\zeta}|y|)(|x|^2+|y|^2)}{4\zeta\sqrt{\zeta}} \right.$$

$$+ \frac{2\cos(\sqrt{\zeta}|x|)\cos(\sqrt{\zeta}|y|)|x||y|}{4\zeta\sqrt{\zeta}}$$

$$\left. - \frac{\sin(\sqrt{\zeta}|x|)\cos(\sqrt{\zeta}|y|)|y|+\cos(\sqrt{\zeta}|x|)\sin(\sqrt{\zeta}|y|)|x|}{2\zeta^2} \right|$$

$$\le \frac{C(1+|x|^2)(1+|y|^2)}{\zeta\sqrt{\zeta}}.$$

For $y < 0$ an identical calculation leads to the same bound. Therefore the operator valued function $\mathcal{F}(\zeta) = \Gamma^+_{11}(\lambda+0)-\Gamma^+_{11}(\lambda-0)$ satisfies the conditions

(4.5.48) and (4.5.49) of Lemma 4.5.16 with $\zeta = -\omega - i\lambda$, $p = 1/2$ and $\mathbf{B} = \mathcal{B}$. □

Next we consider the second summand on the RHS of (4.5.52).

Lemma 4.5.19 *In the situation of Theorem 4.5.17*

$$\int_{\mathcal{C}_+ \cup \mathcal{C}_-} e^{\lambda t} (P(\lambda + 0) - P(\lambda - 0)) \, d\lambda = \mathcal{O}(t^{-3/2}) \qquad (4.5.55)$$

in the norm $\mathcal{B}$.

Proof We consider only the integral over $\mathcal{C}_+$ and one component of the matrix P, for example, P_{11}:

$P_{11}(\lambda, x, y)$

$$= \frac{(i\alpha - 2k_-)e^{ik_+(|x|+|y|)} + (-i\alpha + 2k_+)e^{ik_-(|x|+|y|)} + i\beta(e^{ik_-|y|+ik_+|x|} - e^{ik_+|y|+ik_-|x|})}{2i\alpha(k_+ + k_-) - 4k_+k_- + \alpha^2 - \beta^2}.$$

Denote $\zeta = -\omega - i\lambda$, then $k_+ = \sqrt{\zeta}$, $k_- = \sqrt{-2\omega - \zeta}$. A Taylor expansion in $\sqrt{\zeta}$ as $\zeta \to 0$, Im $\zeta \geq 0$ implies

$$P_{11}(\lambda, x, y) = P_0 + P_1(x, y)\zeta^{1/2} + P_2(x, y)\mathcal{O}(\zeta),$$

where $|P_j(x, y)| \leq C_j(1 + |x|^j)(1 + |y|^j)$, $j = 1, 2$. Therefore, if $\lambda \in \mathcal{C}_+$ then

$$\mathcal{F}(\zeta) = P_{11}(\lambda + 0) - P_{11}(\lambda - 0) = \mathcal{O}(\zeta^{1/2}), \quad \zeta \to 0$$

in the norm of $\mathcal{B}$. Similarly, differentiating two times the function $P_{11}(\lambda, x, y)$ in λ, we obtain that

$$\mathcal{F}''(\zeta) = -P_{11}''(\lambda + 0) + P_{11}''(\lambda - 0) = \mathcal{O}(\zeta^{-3/2}), \quad i, j = 1, 2, \quad \zeta \to 0$$

in the norm of $\mathcal{B}$. Moreover, $\mathcal{F}''(\zeta) \sim \zeta^{-3/2}$ as $\zeta \to \infty$. Therefore, the function $\mathcal{F}(\zeta)$ satisfies the conditions (4.5.48) and (4.5.49) of Lemma 4.5.16 with $p = 1/2$ and $\mathbf{B} = \mathcal{B}$. □

4.5.8 Bounds for Small Times

As a starting point for the method of majorants in Section 4.5.10 we will need also some estimates on the dynamical group e^{Ct} for small t. First note that the function $e^{Ct}\chi_0$ belongs to $C_b(\mathbb{R}, H^1)$. This follows from a theorem analogous to Theorem 4.5.2 for solutions $e^{Ct}\chi_0$ of the linearized equation (4.5.32), with initial condition $\chi_0 \in H^1$. Moreover, energy and charge conservation imply that

$$\|e^{\mathbf{C}t}\chi_0\|_{H^1} \leq c\|\chi_0\|_{H^1}, \quad t \in \mathbb{R}. \tag{4.5.56}$$

For a further application in Section 4.5.11 we need a bound for the action of $e^{\mathbf{C}t}$ on the Dirac distribution $\delta = \delta(x)$.

Thus let $\chi_\delta(x,t)$ be the solution to the linearized equation (4.5.25) with $\chi_\delta(x,0) = \delta(x)$ and $e^{\mathbf{C}t}\delta$ its real vector version. Note that, by Theorem 4.5.17, we have $e^{\mathbf{C}t}\delta \in C_b(\varepsilon,\infty; L_{-\beta}^\infty)$, for every $\varepsilon > 0$, and $\beta \geq 2$. The next lemma gives the small t behavior:

Lemma 4.5.20 *The following bound holds*

$$\|e^{\mathbf{C}t}\delta\|_{L^\infty} = \mathcal{O}(t^{-1/2}), \quad t \to 0. \tag{4.5.57}$$

Proof By the Duhamel representation for the solution to (4.5.25), we obtain

$$\chi_\delta(x,t) = W_\omega(t)\delta - \int_0^t ds \Big(a\chi_\delta(0,s) + b\mathrm{Re}\,(\chi_\delta(0,s)) \Big) W_\omega(t-s)\delta, \tag{4.5.58}$$

where a and b are defined by (4.5.30), and $W_\omega(t)$ is the dynamical group of the modified Schrödinger equation

$$i\dot{\chi}(x,t) = -\chi''(x,t) + \omega\chi(x,t). \tag{4.5.59}$$

Note that

$$W_\omega(t)\delta = \frac{1}{\sqrt{4\pi it}}\, e^{i\frac{x^2}{4t}-i\omega t}. \tag{4.5.60}$$

Therefore (4.5.58) becomes

$$\chi_\delta(x,t) = \frac{1}{\sqrt{4\pi it}}\, e^{i\frac{x^2}{4t}-i\omega t}$$
$$- \int_0^t \frac{1}{\sqrt{4\pi i(t-s)}}\, e^{i\frac{x^2}{4(t-s)}-i\omega(t-s)}\Big(a\chi_\delta(0,s) + b\mathrm{Re}\,(\chi_\delta(0,s))\Big)ds. \tag{4.5.61}$$

Denote $\varsigma(x,t) = \sqrt{t}\,\chi_\delta(x,t)$. Then

$$\varsigma(x,t) = \frac{1}{\sqrt{4\pi i}}\, e^{i\frac{x^2}{4t}-i\omega t}$$
$$- \sqrt{t}\int_0^t \frac{1}{\sqrt{4\pi i(t-s)s}}\, e^{i\frac{x^2}{4(t-s)}-i\omega(t-s)}\Big(a\varsigma(0,s) + b\mathrm{Re}\,(\varsigma(0,s))\Big)ds. \tag{4.5.62}$$

Therefore,

$$\|\varsigma(t)\|_{L^\infty} \le \frac{1}{2\sqrt{\pi}} + \frac{1}{2}\sqrt{\pi t}(|a| + |b|)$$

$$\times \int_0^t \frac{1}{\pi\sqrt{(t-s)s}}\|\varsigma(s)\|_{L^\infty}ds, \qquad t > 0. \qquad (4.5.63)$$

Since $\int_0^t \dfrac{ds}{\pi\sqrt{(t-s)s}} = 1$, we obtain the bound

$$\|\varsigma(t)\|_{L^\infty} \le \frac{1}{2\sqrt{\pi}}\frac{1}{1 - \frac{1}{2}\sqrt{\pi t}(|a| + |b|)}$$

if t is sufficiently small.   □

4.5.9 Modulation Equations

In this section we present the modulation equations which allow a construction of solutions $\psi(x,t)$ of the equation (4.5.1) close at each time t to a soliton, i.e., to one of the functions

$$Ce^{i\theta - \sqrt{\omega}|x|}, \qquad C = C(\omega) > 0$$

in the set S described in Section 4.5.3 with time varying ("modulating") parameters $(\omega, \theta) = (\omega(t), \theta(t))$. It will be assumed that $\psi(x,t)$ is a given weak solution of (4.5.1) as provided by Theorem 4.5.2, so that the map $t \to \psi(\cdot, t)$ is continuous into $H^1(\mathbb{R})$. The modulation equations follow from the ansatz for the solution which is explained next. Recall that we defined

$$\Phi_\omega(x) \equiv \left(Ce^{-\sqrt{\omega}|x|}, 0\right) = \left(\psi_\omega, 0\right) \qquad (4.5.64)$$

so that $\psi(x,t) = e^{j\theta(t)}\Phi_{\omega(t)}(x)$ is a solution of (4.5.28) if and only if $\dot\theta = \omega$ and $\dot\omega = 0$. Here it is to be understood that $C = C(\omega(t))$ is determined from $\omega(t)$ via (4.5.16). We look for a solution to (4.5.28) in the form

$$\psi(x,t) = e^{j\theta(t)}\left(\Phi_{\omega(t)}(x) + \chi(x,t)\right) = e^{j\theta(t)}\Psi(x,t),$$

$$\Psi(x,t) = \Phi_{\omega(t)}(x) + \chi(x,t). \qquad (4.5.65)$$

Since this is a solution of (4.5.28) as long as $\chi \equiv 0$ and $\dot\theta = \omega$ and $\dot\omega = 0$ it is natural to look for solutions in which χ is small and

$$\theta(t) = \int_0^t \omega(s)ds + \gamma(t)$$

with γ treated perturbatively. Observe that so far this representation is underdetermined since for any $(\omega(t), \theta(t))$ it just amounts to a definition of χ; it is made unique by restricting $\chi(t)$ to lie in the image of the projection operator onto the continuous spectrum $\mathbf{P}_t^c = \mathbf{P}^c(\omega(t))$ or equivalently that

$$\mathbf{P}_t^0 \chi(t) = 0, \quad \mathbf{P}_t^0 = \mathbf{P}^0(\omega(t)) = I - \mathbf{P}^c(\omega(t)) \tag{4.5.66}$$

(the projection operators are as defined in (4.5.35) and (4.5.42)). An equivalent formulation of (4.5.66) is to say that $e^{j\theta}\chi$ is required to lie in the symplectic normal space $N_{\omega(t),\theta(t)}S$. This is equivalent to imposition of the following orthogonality conditions (at each time t):

$$\Omega(\chi(t), T_0(\omega(t))) = 0 = \Omega(\chi(t), T_1(\omega(t))), \tag{4.5.67}$$

where Ω is the symplectic form introduced previously. Writing $\chi(t) = (\chi_1(t), \chi_2(t))$ the orthogonality conditions reduce to

$$\int \chi_1(x,t) C e^{-\sqrt{\omega}|x|}\, dx = 0, \quad \text{and} \quad \int \chi_2(x,t) \partial_\omega (C e^{-\sqrt{\omega}|x|})\, dx = 0. \tag{4.5.68}$$

Now we give a system of *modulation equations* for $\omega(t)$, $\gamma(t)$ which ensure the conditions (4.5.68) are preserved by the time evolution.

Lemma 4.5.21 *(i) Assume given a solution of (4.5.28) with regularity as described in Theorem 4.5.2, which can be written in the form (4.5.65)–(4.5.66) with continuously differentiable $\omega(t)$, $\theta(t)$. Then*

$$\dot{\chi} = \mathbf{C}\chi - \dot{\omega}\partial_\omega\Phi_\omega + \dot{\gamma}j^{-1}(\Phi_\omega + \chi) + \mathbf{Q} \tag{4.5.69}$$

where $\mathbf{Q}(\chi, \omega) = -\delta(x)j^{-1}\big(\mathbf{F}(\Phi_\omega + \chi) - \mathbf{F}(\Phi_\omega) - \mathbf{F}'(\Phi_\omega)\chi\big),$ *and*

$$\dot{\omega} = \frac{\langle \mathbf{P}^0 \mathbf{Q}, \Psi \rangle}{\langle \partial_\omega \Phi_\omega - \partial_\omega \mathbf{P}^0 \chi, \Psi \rangle}, \tag{4.5.70}$$

$$\dot{\gamma} = \frac{\langle j\mathbf{P}^0(\partial_\omega\Phi_\omega - \partial_\omega\mathbf{P}^0\chi), \mathbf{P}^0\mathbf{Q}\rangle}{\langle \partial_\omega\Phi_\omega - \partial_\omega\mathbf{P}^0\chi, \Psi\rangle}, \tag{4.5.71}$$

where $\mathbf{P}^0 = \mathbf{P}^0(\omega(t))$ *is the projection operator defined in (4.5.42) and* $\partial_\omega\mathbf{P}^0 = \partial_\omega\mathbf{P}^0(\omega)$ *evaluated at* $\omega = \omega(t)$.
(ii) Conversely, given ψ a solution of (4.5.28) as in Theorem 4.5.2 and continuously differentiable functions $\omega(t)$, $\theta(t)$ which satisfy (4.5.70)–(4.5.71) then χ defined by (4.5.65) satisfies (4.5.69) and the condition (4.5.66) holds at all times if it holds initially.

Proof This can be proved as in [112, Prop.2.2]. □

It remains to show, for appropriate initial data close to a soliton, that there exist solutions to (4.5.70)–(4.5.71), at least locally. To achieve this observe that if the spectral condition (4.5.18) holds then by Lemma 4.5.6 the denominator appearing on the right-hand side of (4.5.70) and (4.5.71) does not vanish for small $\|\chi\|_{L_\beta^1}$. This is because

$$\langle \partial_\omega \psi_s, \psi_s \rangle = \frac{1}{2} \partial_\omega \int |\psi_\omega|^2 dx \neq 0, \tag{4.5.72}$$

as discussed in Section 4.5.3. This has the consequence that the orthogonality conditions really can be satisfied for small χ because they are equivalent to a locally well-posed set of ordinary differential equations for $t \to (\theta(t), \omega(t))$. This implies the following corollary:

Corollary 4.5.22 *(i) In the situation of (i) in the previous lemma assume that (4.5.72) holds. If $\|\chi\|_{L_\beta^p}$ is sufficiently small for some p, β the right-hand sides of (4.5.70) and (4.5.71) are smooth in θ, ω and there exists continuous $\mathcal{R} = \mathcal{R}(\omega, \chi)$ such that*

$$|\dot{\gamma}(t)| \leq \mathcal{R}|\chi(0, t)|^2, \qquad |\dot{\omega}(t)| \leq \mathcal{R}|\chi(0, t)|^2.$$

(ii) Assume given ψ, a solution of (4.5.28) as in Theorem 4.5.2. If ω_0 satisfies (4.5.72) and $\chi(x, 0) = e^{-j\theta_0}\psi(x, 0) - \Phi_{\omega_0}(x)$ is small in some L_β^p norm and satisfies (4.5.66) there is a time interval on which there exist C^1 functions $t \mapsto (\omega(t), \gamma(t))$ which satisfy (4.5.70)–(4.5.71).

4.5.10 Time Decay for the Transversal Dynamics

In this section we state our Theorem 4.5.24 on the time decay of the transversal component $\chi(t)$ in the nonlinear setting, leaving the proof to the next section. Theorem 4.5.24 will be used to prove the main theorem in Section 4.5.12. First we represent the initial data ψ_0 in a convenient form for application of the modulation equations: the next lemma will allow us to assume that (4.5.66) holds initially without loss of generality.

Lemma 4.5.23 *In the situation of Theorem 4.5.9 there exists a stationary orbit $\psi_{\tilde{\omega}_0} = \tilde{C}_0 e^{-\sqrt{\tilde{\omega}_0}|x|}$ satisfying the spectral condition (4.5.18) such that in vector form*

$$\psi_0(x) = e^{j\tilde{\theta}_0}(\Phi_{\tilde{\omega}_0}(x) + \chi_0(x)), \quad \Phi_{\tilde{\omega}_0} = (\psi_{\tilde{\omega}_0}, 0),$$

and for $\chi_0(x)$ we have

$$\mathbf{P}^0(\tilde{\omega}_0)(\chi_0) = 0, \tag{4.5.73}$$

and

$$\|\chi_0\|_{L^1_\beta \cap H^1} = \tilde{d} = O(d) \quad \text{as } d \to 0.$$

Proof By (4.5.67), the condition (4.5.73) is equivalent to the pair of equations

$$\Omega(e^{-j\tilde{\theta}_0}\psi_0 - \Phi_{\tilde{\omega}_0}, T_0(\tilde{\omega}_0)) = 0, \quad \Omega(e^{-j\tilde{\theta}_0}\psi_0 - \Phi_{\tilde{\omega}_0}, T_1(\tilde{\omega}_0)) = 0,$$

where $T_0(\omega) = j\Phi_\omega$, $T_1(\omega) = \partial_\omega\Phi_\omega$. For ψ_0 sufficiently close (in L^1_β) to $e^{j\theta_0}\Phi_{\omega_0}$, the existence of $\tilde{\theta}_0, \tilde{\omega}_0$ follows by a standard application of the implicit function theorem if we show that the Jacobian matrix

$$\begin{pmatrix} \partial_\omega\Omega(e^{-j\theta}\psi_0 - \Phi_\omega, j\Phi_\omega) & \partial_\omega\Omega(e^{-j\theta}\psi_0 - \Phi_\omega, \partial_\omega\Phi_\omega) \\ \partial_\theta\Omega(e^{-j\theta}\psi_0 - \Phi_\omega, j\Phi_\omega) & \partial_\theta\Omega(e^{-j\theta}\psi_0 - \Phi_\omega, \partial_\omega\Phi_\omega) \end{pmatrix}, \quad (4.5.74)$$

with $\psi_0 = e^{j\theta_0}\Phi_{\omega_0}$, is nondegenerate at $\omega = \omega_0$ and $\theta = \theta_0$. But this is equivalent to the nondegeneracy of the matrix

$$\begin{pmatrix} \Omega(\partial_\omega\Phi_{\omega_0}, j\Phi_{\omega_0}) & 0 \\ 0 & \Omega(j\Phi_{\omega_0}, \partial_\omega\Phi_{\omega_0}) \end{pmatrix}, \quad (4.5.75)$$

which holds by (4.5.72). $\qquad\square$

In Section 4.5.12 we will show that our main Theorem 4.5.9 can be derived from the following time decay of the transversal component $\chi(t)$:

Theorem 4.5.24 *Let all the assumptions of Theorem 4.5.9 hold. For d sufficiently small there exist C^1 functions $t \mapsto (\omega(t), \gamma(t))$ defined for $t \geq 0$ such that the solution $\psi(x,t)$ of (4.5.28) can be written as in (4.5.65)–(4.5.66) with (4.5.70)–(4.5.71) satisfied, and there exists a number $\overline{M} > 0$, depending only on the initial data, such that*

$$M(T) = \sup_{0 \leq t \leq T} [(1+t)^{3/2}\|\chi(t)\|_{L^\infty_{-\beta}} + (1+t)^3(|\dot{\gamma}| + |\dot{\omega}|)] \leq \overline{M}, \quad (4.5.76)$$

uniformly in $T > 0$, and $\overline{M} = O(d)$ as $d \to 0$.

Remarks 4.5.25 (i) This theorem will be deduced from Proposition 4.5.26 in the next section.

(ii) Theorem 4.5.2 implies that the norms in the definition of M are continuous functions of time (and so M is also).

(iii) The result holds also for negative time with appropriate changes since $\psi(x,t)$ solves (4.5.1) if and only if $\overline{\psi}(x, -t)$ does.

(iv) The result implies in particular that $t^3|\dot{\theta} - \omega| + t^3|\dot{\omega}| \leq C$, hence $\omega(t)$ and $\theta(t) - t\omega_+$ should converge as $t \to \infty$ while $\psi(x,t) - e^{j\theta(t)}\Phi_{\omega(t)}(x)$ have limit zero in $L^\infty_{-\beta}(\mathbb{R})$.

4.5.11 Decay in Transversal Directions

In this section we prove Theorem 4.5.24. Let us write the initial data in the form

$$\psi_0(x) = e^{j\theta_0}(\Phi_{\omega_0}(x) + \chi_0(x)), \qquad (4.5.77)$$

with $d = \|\chi_0\|_{L^1_\beta \cap H^1}$ sufficiently small. By Lemma 4.5.23 we can assume that $\mathbf{P}^0(\omega_0)(\chi_0) = 0$ without loss of generality. Then the local existence asserted in Corollary 4.5.22 implies the existence of an interval $[0, t_1]$ on which are defined C^1 functions $t \mapsto (\omega(t), \gamma(t))$ satisfying (4.5.70)–(4.5.71) and such that $M(t_1) = \rho$ for some $t_1 > 0$ and $\rho > 0$. By continuity we can make ρ as small as we like by making d and t_1 small.

Proposition 4.5.26 *In the situation of Theorem 4.5.24 let $M(t_1) \leq \rho$ for some $t_1 > 0$ and $\rho > 0$. Then there exist numbers d_1 and ρ_1, independent of t_1, such that*

$$M(t_1) \leq \rho/2 \qquad (4.5.78)$$

if $d = \|\chi_0\|_{L^1_\beta \cap H^1} < d_1$ and $\rho < \rho_1$.

Proof of Theorem 4.5.24 Assuming the truth of Proposition 4.5.26 for now Theorem 4.5.24 will follow from the next argument:

Consider the set $\mathcal{T}$ of $t_1 \geq 0$ such that $(\omega(t), \gamma(t))$ are defined on $[0, t_1]$ and $M(t_1) \leq \rho$. This set is relatively closed by continuity. On the other hand, (4.5.78) and Corollary 4.5.22 with sufficiently small ρ and d imply that this set is also relatively open, and hence $\sup \mathcal{T} = +\infty$, completing the proof of Theorem 4.5.24.

In the remaining part of the section we prove Proposition 4.5.26.

Frozen Linearized Equation

A crucial part of the proof of Proposition 4.5.26 is the estimation of the first term in M, for which purpose it is necessary to make use of the dispersive properties obtained in Sections 4.5.6 and 4.5.7. Rather than study directly (4.5.69), whose linear part is nonautonomous, it is convenient (following [110, 111]) to introduce a second ansatz, a small modification of (4.5.65), which leads to an autonomous linearized equation. This new ansatz for the solution is

$$\psi(x,t) = e^{j\theta}(\Phi_\omega(x) + e^{-j(\theta - \tilde{\theta})}\eta),$$

$$\text{where } \tilde{\theta}(t) = \omega_1 t + \theta_0 \text{ and } \omega_1 = \omega(t_1) \qquad (4.5.79)$$

so that $\eta = e^{j(\theta - \tilde{\theta})}\chi$ and $\chi = e^{-j(\theta - \tilde{\theta})}\eta$. Since

$$\dot{\chi} = e^{-j(\theta - \tilde{\theta})}\big(\dot{\eta} - j(\omega + \dot{\gamma} - \omega_1)\eta\big),$$

equation (4.5.69) implies

$$\dot{\eta} = j^{-1}(\omega_1 - \omega)\eta + e^{j(\theta - \tilde{\theta})}\mathbf{C}\big(e^{-j(\theta - \tilde{\theta})}\eta\big)$$

$$+ e^{j(\theta - \tilde{\theta})}\big(j^{-1}\dot{\gamma}\Phi_\omega - \dot{\omega}\partial_\omega\Phi_\omega + \mathbf{Q}[e^{-j(\theta - \tilde{\theta})}\eta]\big). \qquad (4.5.80)$$

The matrices $\mathbf{C}$ and $e^{j\phi}$, where $\phi = \theta - \tilde{\theta}$, do not commute hence we need the following lemma:

Lemma 4.5.27

$$\mathbf{C}e^{j\phi} - e^{j\phi}\mathbf{C} = \delta(x)b\sin\phi\,\sigma, \text{ where } \sigma = \begin{pmatrix} 1 & 0 \\ 0 & -1 \end{pmatrix}, b = 2a'C^2.$$
$$(4.5.81)$$

Proof

$$\mathbf{C}e^{j\phi} - e^{j\phi}\mathbf{C} = \begin{pmatrix} 0 & \mathbf{D}_2 \\ -\mathbf{D}_1 & 0 \end{pmatrix}\begin{pmatrix} \cos\phi & -\sin\phi \\ \sin\phi & \cos\phi \end{pmatrix}$$

$$- \begin{pmatrix} \cos\phi & -\sin\phi \\ \sin\phi & \cos\phi \end{pmatrix}\begin{pmatrix} 0 & \mathbf{D}_2 \\ -\mathbf{D}_1 & 0 \end{pmatrix}$$

$$= \begin{pmatrix} (\mathbf{D}_2 - \mathbf{D}_1)\sin\phi & 0 \\ 0 & (\mathbf{D}_1 - \mathbf{D}_2)\sin\phi \end{pmatrix}$$

$$= \begin{pmatrix} \delta(x)b\sin\phi & 0 \\ 0 & -\delta(x)b\sin\phi \end{pmatrix}. \qquad \square$$

Using Lemma 4.5.27, we rewrite equation (4.5.80) as

$$\dot{\eta} = j^{-1}(\omega_1 - \omega)\eta + \mathbf{C}\eta$$

$$+ e^{j(\theta - \tilde{\theta})}\Big(-\delta(x)b\sin(\theta - \tilde{\theta})\sigma\eta + j^{-1}\dot{\gamma}\Phi_\omega - \dot{\omega}\partial_\omega\Phi_\omega + \mathbf{Q}[e^{-j(\theta - \tilde{\theta})}\eta]\Big).$$

To obtain a perturbed *autonomous* equation, we rewrite the first two terms on the RHS by freezing the coefficients at $t = t_1$. Note that

$$j^{-1}(\omega_1 - \omega) + \mathbf{C} = \mathbf{C_1} - j^{-1}\delta(x)(V - V_1),$$

where $V = a + bP_1$, $V_1 = V(t_1)$, and $\mathbf{C_1} = \mathbf{C}(t_1)$. The equation for η now reads

$$\dot{\eta} = \mathbf{C_1}\eta - j^{-1}\delta(x)(V - V_1)\eta$$

$$+ e^{j(\theta - \tilde{\theta})}\left(-\delta(x)b\sin(\theta - \tilde{\theta})\sigma\eta + j^{-1}\dot{\gamma}\Phi - \dot{\omega}\partial_\omega\Phi_\omega + \mathbf{Q}[e^{-j(\theta - \tilde{\theta})}\eta]\right).$$

(4.5.82)

The first term is now independent of t; the idea is that if there is sufficiently rapid convergence of $\omega(t)$ as $t \to \infty$ the other remaining terms are small *uniformly with respect to* t_1. Finally, the equation (4.5.82) can be written in the following *frozen form*

$$\dot{\eta} = \mathbf{C_1}\eta + \mathbf{f_1}, \qquad (4.5.83)$$

where

$$\mathbf{f_1} = -j^{-1}\delta(x)(V - V_1)\eta$$

$$+ e^{j(\theta - \tilde{\theta})}\left(-\delta(x)b\sin(\theta - \tilde{\theta})\sigma\eta + j^{-1}\dot{\gamma}\Phi - \dot{\omega}\partial_\omega\Phi_\omega + \mathbf{Q}[e^{-j(\theta - \tilde{\theta})}\eta]\right).$$

(4.5.84)

Remark 4.5.28 The advantage of (4.5.83) over (4.5.69) is that it can be treated as a perturbed autonomous linear equation, so that the estimates from Section 4.5.6 can be used directly. The additional terms in $\mathbf{f_1}$ can be estimated as small uniformly in t_1: see Lemma 4.5.29 below. This is the reason for introduction of the second ansatz (4.5.79).

Lemma 4.5.29 *In the situation of Proposition 4.5.26 there exists $c > 0$, independent of t_1, such that for $0 \leq t \leq t_1$,*

$$|a(t) - a_1| + |b(t) - b_1| + |\theta(t) - \tilde{\theta}(t)| \leq c\rho.$$

Proof By (4.5.76)

$$\sup_{0 \leq t \leq t_1} (1 + t^3)(|\dot{\gamma}(t)| + |\dot{\omega}(t)|) \leq M(t_1) = \rho. \qquad (4.5.85)$$

Therefore

$$|a(t) - a(t_1)| = \left|\int_t^{t_1} \dot{a}(\tau)d\tau\right| \leq c\left(\sup_{0 \leq \tau \leq t_1} (1 + \tau^2)|\dot{\omega}(\tau)|\right)\int_t^{t_1} \frac{d\tau}{1 + \tau^2} \leq c\rho,$$

since $|\dot{a}(\tau)| \leq c|\dot{\omega}(\tau)|$. The difference $|b(t) - b(t_1)|$ can be estimated similarly.

Next,

$$\theta(t) - \tilde{\theta}(t) = \int_0^t \omega(\tau)d\tau + \gamma(t) - \omega(t_1)t - \gamma(0)$$

$$= \int_0^t (\omega(\tau) - \omega(t_1))d\tau + \int_0^t \dot{\gamma}(\tau)d\tau$$

$$= -\int_0^t \int_\tau^{t_1} \dot{\omega}(s)ds\,d\tau + \int_0^t \dot{\gamma}(\tau)d\tau. \qquad (4.5.86)$$

By (4.5.85), the first summand on the RHS of (4.5.86) can be estimated as

$$\int_0^t \int_{t_1}^\tau |\dot{\omega}(s)|ds\,d\tau \le \int_0^t \int_\tau^{t_1} (1+s)^{2+\varepsilon}|\dot{\omega}(s)| \frac{1}{(1+s)^{2+\varepsilon}}ds\,d\tau$$

$$\le c \sup_{0 \le s \le t_1} (1+s)^{2+\varepsilon}|\dot{\omega}(s)| \int_0^t \int_\tau^{t_1} \frac{1}{(1+s)^{2+\varepsilon}}ds\,d\tau \le c\rho$$

since the last integral is bounded for $t \in [0, t_1]$. Finally, for the second summand on the RHS of (4.5.86), inequality (4.5.85) implies

$$\left| \int_0^t \dot{\gamma}(\tau)d\tau \right| \le c \sup_{0 \le \tau \le t_1} (1+\tau^2)|\dot{\gamma}(\tau)| \int_t^{t_1} \frac{d\tau}{1+\tau^2} \le c\rho. \qquad \square$$

Projection onto Discrete and Continuous Spectral Spaces

From Sections 4.5.6 and 4.5.7 we have information concerning $U(t) = e^{\mathbf{C}_1 t}$, in particular, decay on the subspace orthogonal to the (two-dimensional) generalized null space. It is therefore necessary to introduce a further decomposition to take advantage of this. Recall, by comparing (4.5.65) and (4.5.79), that

$$\eta = e^{j(\theta - \tilde{\theta})}\chi \quad \text{and} \quad \mathbf{P}_t^0 \chi(t) = 0. \qquad (4.5.87)$$

Introduce the symplectic projections $\mathbf{P}_1^0 = \mathbf{P}_{t_1}^0$ and $\mathbf{P}_1^c = \mathbf{P}_{t_1}^c$ onto the discrete and continuous spectral subspaces defined by the operator $\mathbf{C}_1$ and write, at each time $t \in [0, t_1]$:

$$\eta(t) = g(t) + h(t) \qquad (4.5.88)$$

with $g(t) = \mathbf{P}_1^0 \eta(t)$ and $h(t) = \mathbf{P}_1^c \eta(t)$. The following lemma shows that it is only necessary to estimate $h(t)$.

Lemma 4.5.30 *In the situation of Proposition 4.5.26, assume*

$$\sup_{0 \le t \le t_1} \left(|\omega(t) - \omega_1| + |\theta(t) - \theta_1(t)| \right) = \Delta$$

is sufficiently small. Then for $0 \le t \le t_1$ there exists a linear operator $\Xi(t)$, bounded on $L_{-\beta}^\infty \cap H^1$, and $c(\Delta, \omega_1)$ such that $\eta(t) = \Xi(t) h(t)$, and

$$c(\Delta, \omega_1)^{-1} \|h\|_{L_{-\beta}^\infty \cap H^1} \le \|\eta\|_{L_{-\beta}^\infty \cap H^1} \le c(\Delta, \omega_1) \|h\|_{L_{-\beta}^\infty \cap H^1}. \quad (4.5.89)$$

Proof Explicitly we write

$$\eta(t) = h(t) + g(t), \qquad g(t) = b_0(t) T_0(\omega_1) + b_1(t) T_1(\omega_1), \quad (4.5.90)$$

where b_0, b_1 are chosen at each time t to ensure that $\Omega(h(t), T_0(\omega_1)) = \Omega(h(t), T_1(\omega_1)) = 0$. Using the fact that (since $\mathbf{P}_t^0 \chi(t) = 0$)

$$\Omega\left(e^{-j(\theta - \tilde{\theta})} \eta, T_0(\omega(t))\right) = 0 = \Omega\left(e^{-j(\theta - \tilde{\theta})} \eta, T_1(\omega(t))\right),$$

this means that b_0, b_1 are determined by

$$-\boldsymbol{\mu}_{\omega_1} b_0(t) = \Omega\big(\eta(t), T_1(\omega_1)\big) = \Omega\big(\eta(t), T_1(\omega_1) - e^{j(\theta - \tilde{\theta})} T_1(\omega(t))\big),$$
$$(4.5.91)$$

$$\boldsymbol{\mu}_{\omega_1} b_1(t) = \Omega\big(\eta(t), T_0(\omega_1)\big) = \Omega\big(\eta(t), T_0(\omega_1) - e^{j(\theta - \tilde{\theta})} T_0(\omega(t))\big).$$
$$(4.5.92)$$

From these it follows that there exists $c > 0$ such that $\|g(t)\|_{L_{-\beta}^\infty \cap H^1} \le c\Delta \|\eta(t)\|_{L_{-\beta}^\infty \cap H^1}$ and hence (4.5.89) follows as claimed. $\qquad \square$

Proof of Proposition 4.5.26

To prove Proposition 4.5.26, we explain how to estimate both terms in M (4.5.76) to be $\le \rho/4$, uniformly in t_1.

Estimation of the second term in M. As in Corollary 4.5.22, we have

$$|\dot{\gamma}(t)| + |\dot{\omega}(t)| \le c_0 |\chi(0,t)|^2 \le c_0 \frac{M(t)^2}{(1 + |t|)^3}, \quad t \le t_1,$$

since $|\chi(0,t)| \le \|\chi(t)\|_{L_{-\beta}^\infty}$. Finally let $\rho_1 < 1/(4c_0)$ to complete the estimate for the second term in M as $\le \rho/4$.

Estimation of the first term in M. By Lemma 4.5.30 it is enough to estimate h. Let us apply the projection $\mathbf{P}_1^c$ to both sides of (4.5.83). Then the equation for h reads

$$\dot{h} = \mathbf{C}_1 h + \mathbf{P}_1^c \mathbf{f}_1. \quad (4.5.93)$$

Now, to estimate h we use the Duhamel representation:

$$h(t) = U(t)h(0) + \int_0^t U(t-s)\mathbf{P}_1^c \mathbf{f_1}(s)ds, \quad t \le t_1 \qquad (4.5.94)$$

with $U(t) = e^{\mathbf{C}_1 t}$ the one parameter group just introduced. Recall that $\mathbf{P}_1^0 h(t) = 0$ for $t \in [0, t_1]$. Therefore

$$\|U(t)h(0)\|_{L_{-\beta}^\infty} \le c(1+t)^{-3/2}\|h(0)\|_{L_\beta^1 \cap H^1} \le c(1+t)^{-3/2}\|\eta(0)\|_{L_\beta^1 \cap H^1} \qquad (4.5.95)$$

by Theorem 4.5.17 and inequalities (4.5.56) and (4.5.89). Let us estimate the integrand on the right-hand side of (4.5.94).

Lemma 4.5.31 *The integrand in (4.5.94) satisfies the following bound: for $0 < s < t$,*

$$\|U(t-s)\mathbf{P}_1^c \mathbf{f_1}(\mathbf{s})\|_{L_{-\beta}^\infty} \le c\frac{1}{(t-s)^{1/2}(1+t-s)}\left(\|\eta(s)\|_{L_{-\beta}^\infty}^2 + \rho\|\eta(s)\|_{L_{-\beta}^\infty}\right). \qquad (4.5.96)$$

Proof We consider two different cases : $t - s > \nu$, and $0 < t - s < \nu$, where $\nu = \nu(a, b)$ is defined in Lemma 4.5.20.

(i) $t - s > \nu$: We use the representation (4.5.84) for $\mathbf{f_1}$ and apply Theorem 4.5.17, Corollary 4.5.22, and Lemma 4.5.29 to obtain that for $t \le t_1$,

$$\begin{aligned}
\|U(t-s)\mathbf{P}_1^c \mathbf{f_1}\|_{L_{-\beta}^\infty} &\le c(\nu)(1+t-s)^{-3/2}\|\mathbf{P}_1^c(\mathbf{f_1}(t))\|_{\mathcal{M}_\beta} \\
&\le c(\nu)(1+t-s)^{-3/2}\left(|\eta(0,t)|^2 + \rho|\eta(0,t)|\right) \\
&\le c(\nu)(1+t-s)^{-3/2}\left(\|\eta(t)\|_{L_{-\beta}^\infty}^2 + \rho\|\eta(t)\|_{L_{-\beta}^\infty}\right).
\end{aligned} \qquad (4.5.97)$$

(ii) $0 < t - s < \nu$: We denote $\tilde{\mathbf{Q}} = \delta(x)\mathbf{Q}$, and represent $\mathbf{f_1}(x,s)$ as

$$\mathbf{f_1}(x,s) = p(s)\delta(x) + q(s)\Phi_\omega + r(s)\partial_\omega \Phi_\omega, \qquad (4.5.98)$$

where

$$\begin{aligned}
p(s) = &-j^{-1}(V - V_1)\eta(0,s) \\
&+ e^{j(\theta-\tilde\theta)}\left(b\sin(\theta - \tilde\theta)\sigma\eta(0,s) + \tilde{\mathbf{Q}}[e^{-j(\theta-\tilde\theta)}\eta(0,s)]\right)
\end{aligned}$$

is an $\mathbb{R}^2$ valued function of time, and

$$q(s) = -e^{-j(\theta-\tilde\theta)}j\dot\gamma, \quad r(s) = -e^{-j(\theta-\tilde\theta)}\dot\omega$$

are (2×2) matrix valued functions of time. Writing $\| \cdot \|$ for both the standard Euclidean and operator norms on these, we have, by Lemma 4.5.29,

$$\|p(s)\| \leq c\big(|\eta(0,s)|^2 + \rho|\eta(0,s)|\big) \leq c\big(\|\eta(s)\|_{L^\infty_{-\beta}}^2 + \rho\|\eta(s)\|_{L^\infty_{-\beta}}\big)$$

and by Corollary 4.5.22,

$$\|q(s)\|, \|r(s)\| \leq c|\eta(0,s)|^2 \leq c\|\eta(s)\|_{L^\infty_{-\beta}}^2.$$

Applying projector $\mathbf{P}_1^c$ to $\mathbf{f_1}$, we obtain

$$\mathbf{P}_1^c\mathbf{f_1}(x,s) = p(s)\delta(x) + q(s)\Phi_\omega + r(s)\partial_\omega\Phi_\omega - \mathbf{P}_1^0\mathbf{f_1}(x,s). \qquad (4.5.99)$$

By Lemma 4.5.20, for sufficiently small ν we obtain

$$\|U(t-s)p(s)\delta(x)\|_{L^\infty_{-\beta}}$$
$$\leq \|U(t-s)p(s)\delta(x)\|_{L^\infty} \leq c(\nu)\|p(s)\|(t-s)^{-1/2}$$
$$\leq c(\nu)(t-s)^{-1/2}\big(\|\eta(s)\|_{L^\infty_{-\beta}}^2 + \rho\|\eta(s)\|_{L^\infty_{-\beta}}\big), \quad 0 < t-s < \nu.$$
$$(4.5.100)$$

By inequality (4.5.56), we have

$$\|U(t-s)\big(q(s)\Phi_\omega + r(s)\partial_\omega\Phi_\omega\big)\|_{L^\infty_{-\beta}}$$
$$\leq c\|U(t-s)\big(q(s)\Phi_\omega + r(s)\partial_\omega\Phi_\omega\big)\|_{H^1}$$
$$\leq c\big(\|q(s)\|\|\Phi_\omega\|_{H^1} + \|r(s)\|\|\partial_\omega\Phi_\omega\|_{H^1}\big)$$
$$\leq c\|\eta(s)\|_{L^\infty_{-\beta}}^2, \quad 0 \leq t-s < \nu. \qquad (4.5.101)$$

The definition (4.5.42) of the projector $\mathbf{P}_1^0$ implies immediately that

$$\|\mathbf{P}_1^0\mathbf{f_1}\|_{H^1} \leq c\big(\|p(s)\| + \|q(s)\| + \|r(s)\|\big).$$

Then, similarly to (4.5.101), we obtain

$$\|U(t-s)\mathbf{P}_1^0\mathbf{f_1}\|_{L^\infty_{-\beta}} \leq c\big(\|\eta(s)\|_{L^\infty_{-\beta}}^2 + \rho\|\eta(s)\|_{L^\infty_{-\beta}}\big), \quad 0 \leq t-s < \nu.$$
$$(4.5.102)$$

Finally, (4.5.99)–(4.5.102) imply

$$\|U(t-s)\mathbf{P}_1^c\mathbf{f_1}\|_{L^\infty_{-\beta}}$$
$$\leq c(t-s)^{-1/2}\big(\|\eta(s)\|_{L^\infty_{-\beta}}^2 + \rho\|\eta(s)\|_{L^\infty_{-\beta}}\big), \quad 0 < t-s < \nu. \quad (4.5.103)$$

From (4.5.97) and (4.5.103), inequality (4.5.96) follows. $\qquad\qquad\square$

Now (4.5.89), (4.5.94), (4.5.95), and (4.5.96) imply

$$\|\eta(t)\|_{L^\infty_{-\beta}} \le c(1+t)^{-3/2}\|\eta(0)\|_{L^1_\beta \cap H^1}$$

$$+ c_1 \int_0^t \frac{ds}{(t-s)^{1/2}(1+t-s)}\left(\|\eta(s)\|^2_{L^\infty_{-\beta}} + \rho\|\eta(s)\|_{L^\infty_{-\beta}}\right).$$

Multiply by $(1+t)^{3/2}$ to deduce

$$(1+t)^{3/2}\|\eta(t)\|_{L^\infty_{-\beta}} \le cd + c_1 \int_0^t \frac{(1+t)^{3/2}(1+s)^{-3}}{(t-s)^{1/2}(1+t-s)}(1+s)^3\|\eta(s)\|^2_{L^\infty_{-\beta}}ds$$

$$+ c_1\rho \int_0^t \frac{(1+t)^{3/2}(1+s)^{-3/2}}{(t-s)^{1/2}(1+t-s)}(1+s)^{3/2}\|\eta(s)\|_{L^\infty_{-\beta}}ds$$

$$(4.5.104)$$

since $\|\eta(0)\|_{L^1_\beta \cap H^1} \le d$. Introduce the majorant

$$m(t) := \sup_{[0,t]}(1+s)^{3/2}\|\eta(s)\|_{L^\infty_{-\beta}}, \quad t \le t_1$$

and hence

$$m(t) \le cd + c_1 m^2(t) \int_0^t \frac{(1+t)^{3/2}(1+s)^{-3}}{(t-s)^{1/2}(1+t-s)}ds$$

$$+ \rho c_1 m(t) \int_0^t \frac{(1+t)^{3/2}(1+s)^{-3/2}}{(t-s)^{1/2}(1+t-s)}ds. \quad (4.5.105)$$

It easy to see (by splitting up the integrals into $s < t/2$ and $s \ge t/2$) that both these integrals are bounded independent of t. Thus (4.5.105) implies that there exist c, c_2, c_3, independent of t_1, such that

$$m(t) \le cd + \rho c_2 m(t) + c_3 m^2(t), \quad t \le t_1.$$

Recall that $m(t_1) \le \rho \le \rho_1$ by assumption. Therefore this inequality implies that $m(t)$ is bounded for $t \le t_1$, and moreover,

$$m(t) \le c_4 d, \quad t \le t_1$$

if d and ρ are sufficiently small. The constant c_4 does not depend on t_1. We choose d in (4.5.19) small enough that $d < \rho/(4c_4)$. Therefore,

$$\sup_{[0,t_1]}(1+t)^{3/2}\|\eta(t)\|_{L^\infty_{-\beta}} < \rho/4$$

if d and ρ are sufficiently small. This bounds the first term as $< \rho/4$ by (4.5.87) and hence $M(t_1) < \rho/2$, completing the proof of Proposition 4.5.26.

4.5.12 Soliton Asymptotics

Here we prove our main Theorem 4.5.9 using the bound (4.5.76). For the solution $\psi(x,t)$ to (4.5.1) let us define the accompanying soliton as $s(x,t) = \psi_{\omega(t)}(x)e^{i\theta(t)}$, where $\dot{\theta}(t) = \omega(t) + \dot{\gamma}(t)$. Then for the difference $z(x,t) = \psi(x,t) - s(x,t)$ we obtain easily from equations (4.5.1) and (4.5.11)

$$
i\dot{z}(x,t) = -z''(x,t) + \dot{\gamma}s(x,t) - i\dot{\omega}\partial_{\omega}s(x,t)
$$

$$
- \delta(x)\big(F(\psi(x,t)) - F(s(x,t))\big). \qquad (4.5.106)
$$

Then

$$
z(t) = W(t)z(0) + \int_0^t W(t-\tau)\big[\dot{\gamma}s(\tau) - i\dot{\omega}\partial_{\omega}s(\tau)
$$

$$
- \delta(x)\big(F(\psi(0,\tau)) - F(s(0,\tau))\big)\big]d\tau, \qquad (4.5.107)
$$

where $z(t) = z(\cdot,t)$, $s(t) = s(\cdot,t)$, and $W(t)$ is the dynamical group of the free Schrödinger equation. Since $\gamma(t) - \gamma_+$, $\omega(t) - \omega_+ = \mathcal{O}(t^{-2})$, and therefore $\theta(t) - \omega_+ t - \gamma_+ = \mathcal{O}(t^{-1})$ for $t \to \infty$, to establish the asymptotic behavior (4.5.20) it suffices to prove that

$$
z(t) = W(t)\Phi_+ + r_+(t) \qquad (4.5.108)
$$

with some $\Phi_+ \in C_b(\mathbb{R}) \cap L^2(\mathbb{R})$ and $\|r_+(t)\|_{C_b(\mathbb{R})\cap L^2(\mathbb{R})} = \mathcal{O}(t^{-1/2})$. Denote $g(t) = \dot{\gamma}s(t) - i\dot{\omega}\partial_{\omega}s(t)$, $h(t) = F(\psi(0,t)) - F(s(0,t))$ and rewrite equation (4.5.107) as

$$
z(t) = W(t)z(0) + W(t)\int_0^t W(-\tau)g(\tau)d\tau - W(t)\int_0^t W(-\tau)\delta(x)h(\tau)d\tau.
$$

$$
\qquad (4.5.109)
$$

Let us recall that $\dot{\omega}(t)$, $\dot{\gamma}(t) \sim t^{-3}$ as $t \to \infty$. Hence, for the second summand in the RHS of (4.5.109) we have

$$
W(t)\int_0^t W(-\tau)g(\tau)d\tau = W(t)\int_0^\infty W(-\tau)g(\tau)d\tau - W(t)\int_t^\infty W(-\tau)g(\tau)d\tau
$$

$$
= W(t)\phi_1 + r_1(t), \qquad (4.5.110)
$$

where, from the unitarity in H^1 of the dynamical group $W(t)$ and the t^{-3} decay of $\dot\omega$ and $\dot\gamma$, we infer that $\phi_1 = \int_0^\infty W(-\tau)g(\tau)d\tau \in H^1$, and $\|r_1(t)\|_{H^1} = \mathcal{O}(t^{-2})$, $t \to \infty$.

Consider now the last summand on the RHS of (4.5.109). Note that $W(t)\delta(x) = \frac{e^{ix^2/(4t)}}{\sqrt{4\pi it}}$, and $|h(t)| \leq c|\chi(0,t)| \leq c(1+t)^{-3/2}$ by (4.5.76). Therefore

$$W(t)\int_0^t W(-\tau)\delta(x)h(\tau)d\tau$$

$$= W(t)\int_0^\infty \frac{e^{-ix^2/(4\tau)}}{\sqrt{-4\pi i\tau}}h(\tau)d\tau - \int_t^\infty \frac{e^{ix^2/(4(t-\tau))}}{\sqrt{4\pi i(t-\tau)}}h(\tau)d\tau$$

$$= W(t)\phi_2 + r_2(t). \tag{4.5.111}$$

Evidently, $\phi_2 = \int_0^\infty \frac{e^{-ix^2/(4\tau)}}{\sqrt{-4\pi i\tau}} h(\tau)d\tau \in C_b$, and $\|r_2(t)\|_{C_b} = \mathcal{O}(t^{-1})$, $t \to \infty$.

Moreover, $\phi_2 \in L^2$, and $\|r_2(t)\|_{L^2} = \mathcal{O}(t^{-1/2})$, $t \to \infty$. To see that this is indeed true, change variable to $\tau = 1/u$ in the definition to get:

$$\phi_2(x) = \frac{1}{\sqrt{-4\pi i}}\int_0^\infty e^{-iux^2/4}\, \eta(u)\, du, \qquad \eta(u) = h(1/u)/u^{3/2}. \tag{4.5.112}$$

Now $h(t)$ is bounded and it follows from the decay of $h(t)$ that $\eta(u)$ is bounded as $u \to 0$. Therefore $\eta(u)$ is square integrable and so by the Parseval theorem ϕ_2 is square integrable as a function of $y = x^2$, and hence also as a function of x (since $dy = 2xdx$ and ϕ_2 is a bounded continuous function). Next we have $r_2(t) = -W(t)R(t)$ with

$$R(x,t) = \frac{1}{\sqrt{-4\pi i}}\int_0^{1/t} e^{-iux^2/4}\, \eta(u)\, du = \frac{1}{\sqrt{-4\pi i}}F_{u\to x^2/4}\zeta_t(u)\eta(u),$$

where $\zeta_t(u)$ is the characteristic function of the interval $(0, 1/t)$. The function $\eta(u)$ is bounded, hence $\|\zeta_t\eta\|_{L^2} = ct^{-1/2}$ and therefore $\|r_2(t)\|_{L^2} = \mathcal{O}(t^{-1/2})$, $t \to \infty$. To conclude, using (4.5.109), (4.5.110), and (4.5.111) we obtain (4.5.108) with $\phi_+ = z(0) + \phi_1 + \phi_2$ and $r_+(t) = r_1(t) + r_2(t)$. The $t \to -\infty$ case is handled in an identical way.

Now Theorem 4.5.9 is proved.

Remark 4.5.32 The expression (4.5.112) for ϕ_2 as a Fourier transform implies immediately that $|\phi_2|$, and hence $|\Phi_+|$ also, tend to 0 as $|x| \to \infty$ by the Riemann–Lebesgue theorem. This same expression could be used with Zygmund's lemma to obtain more detailed decay properties of ϕ_2 and hence of Φ_+. The decay rate would be determined essentially by the regularity of the function $h(t)$ in addition to the decay rate of the initial data.

4.5.13 The Kernel and Poles of the Resolvent

In this section we calculate the resolvent and its poles.

The Kernel of the Resolvent

The derivation of the time decay of the solution to the linearized equation (4.5.25) in Section 4.5.6 required an analysis of the smoothness and singularities of the resolvent $\mathbf{R}(\lambda)$ and its asymptotics as $\lambda \to \infty$. Here we will construct its matrix integral kernel explicitly

$$\mathbf{R}(\lambda, x, y) = \begin{pmatrix} R_{11}(\lambda, x, y) & R_{12}(\lambda, x, y) \\ R_{21}(\lambda, x, y) & R_{22}(\lambda, x, y) \end{pmatrix}. \qquad (4.5.113)$$

It is the solution to the equation

$$(\mathbf{C} - \lambda)\mathbf{R}(\lambda, x, y) = \delta(x - y) \begin{pmatrix} 1 & 0 \\ 0 & 1 \end{pmatrix}. \qquad (4.5.114)$$

Calculation of first column For the first column $R_I(\lambda, x, y) := \begin{pmatrix} R_{11}(\lambda, x, y) \\ R_{21}(\lambda, x, y) \end{pmatrix}$ of the matrix $\mathbf{R}(\lambda, x, y)$, we obtain

$$(\mathbf{C} - \lambda)R_I(\lambda, x, y) = \delta(x - y) \begin{pmatrix} 1 \\ 0 \end{pmatrix}. \qquad (4.5.115)$$

If $x \neq 0$ and $x \neq y$, (4.5.115) takes the form (cf. (4.5.31), (4.5.32))

$$\begin{pmatrix} -\lambda & D_2 \\ -D_1 & -\lambda \end{pmatrix} R_I(\lambda, x, y) = \begin{pmatrix} -\lambda & -\dfrac{d^2}{dx^2} + \omega \\ \dfrac{d^2}{dx^2} - \omega & -\lambda \end{pmatrix} R_I(\lambda, x, y) = 0.$$

$$\qquad (4.5.116)$$

The general solution is a linear combination of exponential solutions of type $e^{ikx}v$. Substituting into (4.5.116), we get

$$\begin{pmatrix} -\lambda & k^2 + \omega \\ -k^2 - \omega & -\lambda \end{pmatrix} v = 0. \qquad (4.5.117)$$

For nonzero vectors v, the determinant of the matrix vanishes,

$$\lambda^2 + (k^2 + \omega)^2 = 0. \tag{4.5.118}$$

Then $k_\pm^2 + \omega = \mp i\lambda$. Finally, we obtain four roots $\pm k_\pm(\lambda)$ with

$$k_\pm(\lambda) = \sqrt{-\omega \mp i\lambda}, \tag{4.5.119}$$

where the square root is defined as an analytic continuation from a neighborhood of the zero point $\lambda = 0$ taking the positive value of Im $\sqrt{-\omega}$ at $\lambda = 0$. We choose the cuts in the complex plane λ from the branching points to infinity: the cut $\mathcal{C}_+ := [i\omega, i\infty)$ for $k_+(\lambda)$ and the cut $\mathcal{C}_- := [-i\omega, -i\infty)$ for $k_-(\lambda)$. Then

$$\text{Im } k_\pm(\lambda) > 0, \qquad \lambda \in \mathbb{C} \setminus \mathcal{C}_\pm. \tag{4.5.120}$$

It remains to derive the vector $v = (v_1, v_2)$ which is the solution to (4.5.117):

$$v_2 = -\frac{k_\pm^2 + \omega}{\lambda} v_1 = \frac{\pm i\lambda}{\lambda} v_1 = \pm i v_1.$$

Therefore, we have two corresponding vectors $v_\pm = \begin{pmatrix} 1 \\ \pm i \end{pmatrix}$ and we get four linearly independent exponential solutions

$$v_+ e^{\pm i k_+ x} = \begin{pmatrix} 1 \\ i \end{pmatrix} e^{\pm i k_+ x}, \qquad v_- e^{\pm i k_- x} = \begin{pmatrix} 1 \\ -i \end{pmatrix} e^{\pm i k_- x}.$$

Now we can solve the equation (4.5.115). First, we rewrite it using the representations (4.5.32) and (4.5.31) for the operator $\mathbf{C}$,

$$\begin{pmatrix} -\lambda & -\frac{d^2}{dx^2} + \omega \\ \frac{d^2}{dx^2} - \omega & -\lambda \end{pmatrix} \begin{pmatrix} R_{11}(\lambda, x, y) \\ R_{21}(\lambda, x, y) \end{pmatrix}$$

$$= \delta(x-y) \begin{pmatrix} 1 \\ 0 \end{pmatrix} + \delta(x) \begin{pmatrix} 0 & a \\ -a - b & 0 \end{pmatrix} \begin{pmatrix} R_{11}(\lambda, 0, y) \\ R_{21}(\lambda, 0, y) \end{pmatrix}.$$

Let us consider $y > 0$ for the sake of concreteness. Then the RHS vanishes in the open intervals $(-\infty, 0), (0, y)$, and (y, ∞). Hence, for the parameter λ outside the cuts $\mathcal{C}_\pm$, the solution admits the representation

$R_I(\lambda, x, y)$
$$= \begin{cases} A_+ e^{-ik_+ x} v_+ + A_- e^{-ik_- x} v_-, & x < 0, \\ B_+^- e^{-ik_+ x} v_+ + B_-^- e^{-ik_- x} v_- + B_+^+ e^{ik_+ x} v_+ + B_-^+ e^{ik_- x} v_-, & 0 < x < y, \\ C_+ e^{ik_+ x} v_+ + C_- e^{ik_- x} v_-, & x > y. \end{cases}$$

since by (4.5.120), the exponent $e^{-ik_\pm x}$ decays for $x \to -\infty$, and similarly, $e^{ik_\pm x}$ decays for $x \to \infty$. Next we need eight equations to calculate the eight constants $A_+, \ldots, C_-$. We have two continuity equations and two jump conditions for the derivatives at the points $x = 0$ and $x = y$. These four vector equations give just eight scalar equations for the calculation.

Continuity at $x = y$: $R_I(y - 0, y) = R_I(y + 0, y)$, i.e.,

$$B_-^- v_+/e_+ + B_-^- v_-/e_- + B_+^+ v_+ e_+ + B_-^+ v_- e_- = C_+ v_+ e_+ + C_- v_- e_-,$$

where $e_\pm := e^{ik_\pm y}$. It is equivalent to

$$(4.5.121)\qquad \begin{cases} B_+^-/e_+ + B_+^+ e_+ = C_+ e_+, \\ B_-^-/e_- + B_-^+ e_- = C_- e_-. \end{cases}$$

Continuity at $x = 0$: $R_I(-0, y) = R_I(+0, y)$, i.e.,

$$A_+ v_+ + A_- v_- = B_+^- v_+ + B_-^- v_- + B_+^+ v_+ + B_-^+ v_-$$

that is equivalent to

$$(4.5.122)\qquad \begin{cases} A_+ = B_+^- + B_+^+, \\ A_- = B_-^- + B_-^+. \end{cases}$$

Jump at $x = y$: $R_I'(y + 0, y) = R_I'(y - 0, y) + \left(\begin{smallmatrix}0\\-1\end{smallmatrix}\right)$, where prime denotes the derivative in x. Substituting (4.5.121), we get

$$ik_+ C_+ v_+ e_+ + ik_- C_- v_- e_- = -ik_+ B_+^- v_+/e_+ - ik_- B_-^- v_-/e_-$$

$$(4.5.123)\qquad + ik_+ B_+^+ v_+ e_+ + ik_- B_-^+ v_- e_- + \left(\begin{matrix}0\\-1\end{matrix}\right).$$

Noting that

$$\left(\begin{matrix}0\\-1\end{matrix}\right) = \frac{v_+ - v_-}{2} i, \qquad (4.5.124)$$

we get

$$(4.5.125)\qquad \begin{cases} ik_+ C_+ e_+ = -ik_+ B_+^-/e_+ + ik_+ B_+^+ e_+ + \dfrac{i}{2}, \\ ik_- C_- e_- = -ik_- B_-^-/e_- + ik_- B_-^+ e_- - \dfrac{i}{2}. \end{cases}$$

After substituting of $C_\pm$ from (4.5.121), the constants $B_\pm^+$ cancel and we get

$$B_+^- = \frac{e_+}{4k_+}, \qquad B_-^- = -\frac{e_-}{4k_-}. \qquad (4.5.126)$$

Jump at $x = 0$: $R_I'(+0, y) = R_I'(-0, y) - \begin{pmatrix} a+b & 0 \\ 0 & a \end{pmatrix} R_I(-0, y)$. Substituting (4.5.121), we get

$$- ik_+ B_+^- v_+ - ik_- B_-^- v_- + ik_+ B_+^+ v_+ + ik_- B_-^+ v_-$$

$$= -ik_+ A_+ v_+ - ik_- A_- v_- - M(A_+ v_+ + A_- v_-), \qquad (4.5.127)$$

where M is the matrix $\begin{pmatrix} a+b & 0 \\ 0 & a \end{pmatrix}$. Note that

$$\begin{cases} Mv_+ = \alpha v_+ + \beta v_- \\ Mv_- = \alpha v_- + \beta v_+ \end{cases}, \quad \text{where} \quad \alpha = a + \frac{b}{2}, \quad \beta = \frac{b}{2}. \qquad (4.5.128)$$

Then (4.5.127) becomes

$$\begin{cases} -ik_+ B_+^- + ik_+ B_+^+ = -ik_+ A_+ - A_+ \alpha - A_- \beta, \\ -ik_- B_-^- + ik_- B_-^+ = -ik_- A_- - A_+ \beta - A_- \alpha. \end{cases}$$

Substituting here (4.5.122), we get after cancellations,

$$\begin{cases} (2ik_+ + \alpha) B_+^+ + \beta B_-^+ = -\alpha B_+^- - \beta B_-^-, \\ \beta B_+^+ + (2ik_- + \alpha) B_-^+ = -\beta B_+^- - \alpha B_-^-. \end{cases}$$

Hence, the solution is given by

$$\begin{pmatrix} B_+^+ \\ B_-^+ \end{pmatrix} = -\frac{1}{D} \begin{pmatrix} 2ik_- + \alpha & -\beta \\ -\beta & 2ik_+ + \alpha \end{pmatrix} \begin{pmatrix} \alpha & \beta \\ \beta & \alpha \end{pmatrix} \begin{pmatrix} B_+^- \\ B_-^- \end{pmatrix},$$
$$(4.5.129)$$

where D is the determinant

$$D := (2ik_+ + \alpha)(2ik_- + \alpha) - \beta^2, \qquad (4.5.130)$$

and B_+^-, B_-^- are given by (4.5.126). The formulas (4.5.126) and (4.5.129) imply

$$\begin{cases} B_+^+ = \frac{1}{2D} \left(-\dfrac{2ik_- \alpha + \alpha^2 - \beta^2}{2k_+} e_+ + i\beta e_- \right) \\ B_-^+ = \frac{1}{2D} \left(-i\beta e_+ + \dfrac{2ik_+ \alpha + \alpha^2 - \beta^2}{2k_-} e_- \right) \end{cases}. \qquad (4.5.131)$$

Using the identities

$$2ik_- \alpha + \alpha^2 - \beta^2 = D - 2ik_+ \alpha + 4k_+ k_-,$$
$$2ik_+ \alpha + \alpha^2 - \beta^2 = D - 2ik_- \alpha + 4k_+ k_-,$$

we rewrite (4.5.131) as

$$
\begin{cases}
B_+^+ = -\dfrac{e_+}{4k_+} + \dfrac{1}{2D}\left((i\alpha - 2k_-)e_+ + i\beta e_-\right) \\[2mm]
B_-^+ = -\dfrac{e_-}{4k_-} - \dfrac{1}{2D}\left(i\beta e_+ + (i\alpha - 2k_+)e_-\right)
\end{cases}.
\tag{4.5.132}
$$

Finally, the formulas (4.5.121)–(4.5.122), (4.5.126), and (4.5.132) give the first column $R_I(\lambda, x, y)$ of the resolvent for $y > 0$:

$$
R_I(\lambda, x, y) = \Gamma_I(\lambda, x, y) + P_I(\lambda, x, y),
\tag{4.5.133}
$$

where

$$
\Gamma_I(\lambda, x, y) = \frac{1}{4k_+}(e^{ik_+|x-y|} - e^{ik_+(|x|+|y|)})v_+
$$

$$
- \frac{1}{4k_-}(e^{ik_-|x-y|} - e^{ik_-(|x|+|y|)})v_-,
\tag{4.5.134}
$$

and

$$
P_I(\lambda, x, y) = \frac{1}{2D}\left[\left((i\alpha - 2k_-)e^{ik_+(|x|+|y|)} + i\beta e^{i(k_+|x|+k_-|y|)}\right)v_+ \right.
\tag{4.5.135}
$$

$$
\left. - \left(i\beta e^{i(k_-|x|+k_+|y|)} + (i\alpha - 2k_+)e^{ik_-(|x|+|y|)}\right)v_-\right].
$$

Calculation of second column The second column is given by similar formulas with the vector $\begin{pmatrix} 1 \\ 0 \end{pmatrix}$ instead of $\begin{pmatrix} 0 \\ 1 \end{pmatrix}$ in (4.5.121). Then $\begin{pmatrix} 0 \\ -1 \end{pmatrix}$ in (4.5.123) is changed by $\begin{pmatrix} 1 \\ 0 \end{pmatrix}$. Respectively, (4.5.124) is changed by

$$
\begin{pmatrix} 1 \\ 0 \end{pmatrix} = \frac{v_- + v_+}{2}.
$$

Hence, we have now changed $i/2$ by $1/2$ in the first equation of (4.5.125) and $-i/2$ by $1/2$ in the second one. Respectively, (4.5.126) for the second column reads

$$
B_+^- = -\frac{ie_+}{4k_+}, \qquad B_-^- = -\frac{ie_-}{4k_-}.
$$

Then the second column $R_{II}(\lambda, x, y)$ of the resolvent reads:

$$
R_{II}(\lambda, x, y) = \Gamma_{II}(\lambda, x, y) + P_{II}(\lambda, x, y),
\tag{4.5.136}
$$

where

$$\Gamma_{II}(\lambda, x, y) = -\frac{i}{4k_+}(e^{ik_+|x-y|} - e^{ik_+(|x|+|y|)})v_+$$

$$- \frac{i}{4k_-}(e^{ik_-|x-y|} - e^{ik_-(|x|+|y|)})v_-, \qquad (4.5.137)$$

and

$$P_{II}(\lambda, x, y) = \frac{i}{2D}\Big[\Big(-(i\alpha - 2k_-)e^{ik_+(|x|+|y|)} + i\beta e^{i(k_+|x|+k_-|y|)}\Big)v_+$$

$$+ \Big(i\beta e^{i(k_-|x|+k_+|y|)} - (i\alpha - 2k_+)e^{ik_-(|x|+|y|)}\Big)v_-\Big].$$
$$(4.5.138)$$

Note that if $y < 0$ we get the same formulas.

The Poles of the Resolvent

The poles of the resolvent correspond to the roots of the determinant (4.5.130),

$$D(\lambda) := \alpha^2 + 2i\alpha(k_+ + k_-) - 4k_+k_- - \beta^2 = 0, \qquad (4.5.139)$$

with $k_\pm$ as in (4.5.119)–(4.5.120). Thus $D(\lambda)$ is an analytic function on $\mathbb{C} \setminus \mathcal{C}_- \cup \mathcal{C}_+$. Since there are two possible values for the square roots in $k_\pm$ there is a corresponding four-sheeted function $\tilde{D}(\lambda)$ analytic on a four-sheeted cover of $\mathbb{C}$ which is branched over $\mathcal{C}_-$ and $\mathcal{C}_+$. We call the sheet defined by (4.5.120) the *physical sheet*.

We will reduce the equation (4.5.139) to the solution of two successive quadratic equations. These can be solved explicitly but the process involves squaring and thus actually produces zeros of the function $\tilde{D}(\lambda)$ rather than of $D(\lambda)$. Therefore we will then have to check whether or not the roots do actually lie on the physical sheet.

Step (i).
Denote by $\sigma = k_+ + k_-$. Then

$$\sigma^2 = 2k_+k_- - 2\omega \qquad (4.5.140)$$

by (4.5.119), hence (4.5.139) gives the *first quadratic equation*:

$$\alpha^2 + 2i\alpha\sigma - 2(\sigma^2 + 2\omega) - \beta^2 = 0.$$

Rewrite it as

$$\sigma^2 - i\alpha\sigma = \frac{\alpha^2 - \beta^2}{2} - 2\omega =: \delta. \qquad (4.5.141)$$

Finally,

$$\sigma = \frac{i\alpha}{2} \pm \sqrt{\delta - \frac{\alpha^2}{4}}, \tag{4.5.142}$$

where the root is chosen arbitrarily.

Further let us express the roots in ω. Since $a = 2\sqrt{\omega}, \alpha = a+b/2, \beta = b/2$ then substituting δ from (4.5.141), we obtain

$$\delta - \frac{\alpha^2}{4} = \frac{\alpha^2}{4} - \frac{\beta^2}{2} - 2\omega = \frac{(a+b/2)^2}{4} - \frac{b^2}{8} - \frac{a^2}{2}$$

$$= -\frac{a^2}{4} - \frac{b^2}{16} + \frac{ab}{4} = -\frac{1}{16}(2a-b)^2 < 0.$$

Now (4.5.142) reads

$$\sigma = \frac{i\alpha}{2} \pm \frac{i}{4}(2a - b) = \frac{i}{4}\Big[(2a + b) \pm (2a - b)\Big] = i\gamma_j, \ j = 1, 2, \tag{4.5.143}$$

where $\gamma_j \in \mathbb{R}$, and

$$\gamma_1 = a = a(C^2), \qquad \gamma_2 = b/2 = a'(C^2)C^2. \tag{4.5.144}$$

Step (ii).
It remains to calculate the corresponding spectral parameter λ. First, the quadratic equation (4.5.140) implies by (4.5.143) that

$$4(k_+ k_-)^2 = (2\omega + \sigma^2)^2 = (2\omega - \gamma_j^2)^2, \ j = 1, 2. \tag{4.5.145}$$

On the other hand,

$$k_+ k_- = \sqrt{-\omega + i\lambda}\sqrt{-\omega - i\lambda}, \tag{4.5.146}$$

hence (4.5.145) gives the *second quadratic equation*

$$4(\omega^2 + \lambda^2) = (2\omega - \gamma_j^2)^2.$$

Therefore,

$$\lambda^2 = \frac{(2\omega - \gamma_j^2)^2 - 4\omega^2}{4} = -\frac{\gamma_j^2(4\omega - \gamma_j^2)}{4}.$$

Finally, we obtain four roots

$$\lambda_j = i\frac{\gamma_j}{2}\sqrt{4\omega - \gamma_j^2}, \tag{4.5.147}$$

where $j \in \{1, 2\}$ and the square root can take two opposite values.

Corollary 4.5.33 *The four-sheeted function $\tilde{D}(\lambda)$ has the following roots (zeros):*
(i) $j = 1$ gives $\lambda_1 = 0$ since $4\omega = a^2 = \gamma_1$.
(ii) If $|\gamma_2| < 2\sqrt{\omega}$, then both $j = 2$ roots $\pm i|\lambda_2|$ are pure imaginary.
(iii) If $|\gamma_2| > 2\sqrt{\omega}$, then both $j = 2$ roots $\pm|\lambda_2|$ are real: one positive and one negative.

Remark 4.5.34 Note that a priori we can meet the wrong sign of $\text{Im}\, k_\pm$ squaring (4.5.146), which is why the above calculation yields roots of $\tilde{D}(\lambda)$ rather than the physical branch $D(\lambda)$. Since the formulas (4.5.133)–(4.5.138) involve only $D(\lambda)$ it is important to know which of these are actually roots of $D(\lambda)$ and also to know the multiplicities. This is done in the next two sections.

Discrete Spectrum $\lambda = 0$

In order to check that the roots of $\tilde{D}(\lambda)$ given in Corollary 4.5.33 are actually roots of $D(\lambda)$ it suffices to check explicitly that $D(\lambda)$ vanishes, with the assumption that we are on the physical branch defined by $\text{Im}\, k_\pm > 0$ for $\lambda \in \mathbb{C} \setminus \mathcal{C}_\pm$.

For $j = 1$ we have $\gamma = \gamma_1 = a = 2\sqrt{\omega}$ and then $\lambda_1 = 0$. For $j = 2$ we have $\gamma = \gamma_2 = a'C^2$. If $|\gamma_2| = 2\sqrt{\omega}$ (equivalently $|a'| = a/C^2$) or $\gamma_2 = 0$ (equivalently $a' = 0$), we have $\lambda_2 = 0$.

Let us check that $\lambda = 0$ is a root of $D(\lambda)$:

$$D(0) = \alpha^2 - \beta^2 + 2i\alpha 2i\sqrt{\omega} + 4\omega$$

$$= (a + b/2)^2 - b^2/4 - 2(a + b/2)a + a^2 = 0$$

since $k_\pm = i\sqrt{\omega}$. Now let us compute $D'(\lambda)$:

$$D'(\lambda) = i\alpha\left(\frac{i}{\sqrt{-\omega + i\lambda}} + \frac{-i}{\sqrt{-\omega - i\lambda}}\right)$$

$$-\left(\frac{2i}{\sqrt{-\omega + i\lambda}} \cdot \sqrt{-\omega - i\lambda} + \frac{-2i}{\sqrt{-\omega - i\lambda}} \cdot \sqrt{-\omega + i\lambda}\right).$$

Hence $D'(0) = 0$ and $\lambda = 0$ is the root of $D(\lambda)$ of multiplicity at least 2. Further calculation shows that the Taylor series for D near zero takes the form:

$$D(\lambda) = \left(\frac{1}{\omega} - \frac{b}{4\omega^{3/2}}\right)\lambda^2 + O(\lambda^4). \tag{4.5.148}$$

Therefore $\lambda = 0$ is the root of $D(\lambda)$ of multiplicity 4 if and only if $b = 4\sqrt{\omega}$, i.e., $a' = a/C^2$, and we have proved the following lemma:

Lemma 4.5.35 *If $a' = a/C^2$ then $\lambda = 0$ is a root of the determinant $D(\lambda)$ with multiplicity 4, otherwise $\lambda = 0$ is a root of the determinant $D(\lambda)$ with multiplicity 2.*

Nonzero Discrete Spectrum

Now let us check whether the roots $\lambda = \lambda_2 \neq 0$ corresponding to $\gamma = \gamma_2 \notin \{0, \pm 2\sqrt{\omega}\}$ lie on the physical branch. We analyze two different cases: $0 < |\gamma_2| < 2\sqrt{\omega}$ and $|\gamma_2| > 2\sqrt{\omega}$.

I. The case $0 < |\gamma_2| < 2\sqrt{\omega}$ (equivalently $0 < |a'| < a/C^2$).

Since $4\omega - \gamma_2^2 > 0$, the corresponding roots λ_2 are pure imaginary by (4.5.147). Moreover, $|\lambda_2| \leq \omega$. Indeed, (4.5.147) implies

$$\omega^2 - |\lambda_2|^2 = \omega^2 + \gamma_2^4/4 - \gamma_2^2\omega = (\omega - \gamma_2^2/2)^2 \geq 0.$$

Hence $-\omega \mp i\lambda_2 \leq 0$ and $k_\pm$ are pure imaginary with non-negative imaginary part, that is

$$k_+k_- \leq 0 \text{ and Im } (k_+ + k_-) > 0. \tag{4.5.149}$$

The equations (4.5.145) and (4.5.140) imply

$$|k_+k_-| = \frac{1}{4}|a^2 - 2(a')^2C^4|, \ (k_+ + k_-)^2 = -2\omega + 2k_+k_- = -\frac{a^2}{2} + 2k_+k_-. \tag{4.5.150}$$

In order to obtain k_+k_- and $k_+ + k_-$ from the last two equations we have to divide the set $0 < |a'| < a/C^2$ onto three subsets:

$$(-a/C^2, a/C^2) \setminus \{0\}$$

$$= (-a/C^2, -a/\sqrt{2}C^2] \cup \left((-a/\sqrt{2}C^2, a/\sqrt{2}C^2) \setminus \{0\}\right) \cup \left[\frac{a}{\sqrt{2}C^2}, \frac{a}{C^2}\right).$$

1) First consider the case $a' \in \left[\frac{a}{\sqrt{2}C^2}, \frac{a}{C^2}\right)$. Then (4.5.149) and (4.5.150) imply

$$k_+k_- = \frac{1}{4}(a^2 - 2(a')^2C^4),$$

$$(k_+ + k_-)^2 = -\frac{a^2}{2} + \frac{a^2}{2} - (a')^2C^4 = -(a')^2C^4,$$

$$k_+ + k_- = ia'C^2,$$

and using (4.5.139), we obtain

$$D(\lambda_2) = (a + a'C^2)^2 - (a'C^2)^2 + 2i(a + a'C^2)(k_+ + k_-) - 4k_+ k_-$$

$$= a^2 + 2aa'C^2 - 2(a + a'C^2)a'C^2 - a^2 + 2(a')^2 C^4 = 0.$$

Note that each γ_2 defines two values λ_2 up to factor ± 1. If we replace λ_2 by $-\lambda_2$, k_+ and k_- change places and our calculation remains valid. Therefore, both values of λ_2 are roots of $D(\lambda)$.

2) Further consider $a' \in \left(-\dfrac{a}{C^2}, -\dfrac{a}{\sqrt{2}C^2} \right]$. In this case

$$k_+ k_- = \frac{1}{4}(a^2 - 2(a')^2 C^4), \quad k_+ + k_- = -ia'C^2.$$

Then we have

$$D(\lambda_2) = a^2 + 2aa'C^2 + 2(a + a'C^2)a'C^2 - a^2 + 2(a')^2 C^4$$

$$= 4a'C^2(a + a'C^2) \neq 0$$

since $a' \neq 0$ and $a' \neq -a/C^2$. Therefore in this case both values of λ_2 are not the roots of $D(\lambda)$.

3) Finally consider $0 < |a'| < \dfrac{a}{\sqrt{2}C^2}$. Then (4.5.149)-(4.5.150) imply that

$$k_+ k_- = -\frac{1}{4}(a^2 - 2(a')^2 C^4) < 0,$$

$$(k_+ + k_-)^2 = -a^2 + (a')^2 C^4 < 0,$$

$$k_+ + k_- = i\sqrt{a^2 - (a')^2 C^4}.$$

Then we have

$$D(\lambda_2) = a(a + 2a'C^2) - 2(a + a'C^2)\sqrt{a^2 - (a')^2 C^4} + a^2 - 2(a')^2 C^4.$$
$$(4.5.151)$$

To solve the equation $D(\lambda_2) = 0$ with respect to a', divide the RHS of (4.5.151) by $C^4 \neq 0$ and denote $p = a/C^2 > 0$. Then we get the equation

$$p^2 + pa' - (a')^2 = (p + a')\sqrt{p^2 - (a')^2}, \ 0 < |a'| < p/\sqrt{2}. \quad (4.5.152)$$

Squaring both sides of (4.5.152), we get

$$2(a')^4 - p^2(a')^2 = 0.$$

The equation has no solutions for $0 < |a'| < p/\sqrt{2}$ and hence $D(\lambda_2)$ does not vanish.

Corollary 4.5.36 *(i)* $D(\lambda_2) = 0$ *if* $a' \in \left[\dfrac{a}{\sqrt{2}C^2}, \dfrac{a}{C^2}\right)$.

(ii) $D(\lambda_2) \neq 0$ *if* $a' \in \left(-\dfrac{a}{C^2}, \dfrac{a}{\sqrt{2}C^2}\right) \setminus \{0\}$.

II. The case $|\gamma_2| > 2\sqrt{\omega}$ (equivalently $|a'| > a/C^2$).
Since $4\omega - \gamma_2^2 < 0$, the corresponding roots (4.5.147) are real: $\lambda_2^- < 0 < \lambda_2^+$, $\lambda_2^- = -\lambda_2^+$. It is easy to prove that $k_\pm$ take the form:

$$k_\pm = \pm\mu + i\nu, \quad \nu > 0.$$

Therefore

$$k_+k_- = -\mu^2 - \nu^2 < 0, \quad k_+ + k_- = 2i\nu. \tag{4.5.153}$$

1) First, consider the case $a' > a/C^2$. Then by (4.5.150) and (4.5.153),

$$k_+k_- = \frac{1}{4}(a^2 - 2(a')^2C^4), \quad (k_+ + k_-)^2 = -(a')^2C^4, \quad k_+ + k_- = ia'C^2.$$

Therefore

$$D(\lambda_2) = a(a + 2a'C^2) + 2i(a + a'C^2)(k_+ + k_-) - 4k_+k_-$$
$$= a(a + 2a'C^2) - 2(a + a'C^2)a'C^2 - a^2 + 2(a')^2C^4 = 0$$

and then λ_2 are real roots of $D(\lambda)$. Hence, the case $a' > a/C^2$ is **linearly unstable**.

2) Further consider the case $a' < -a/C^2$. Then

$$k_+k_- = \frac{1}{4}(a^2 - 2(a')^2C^4) < 0, \quad k_+ + k_- = -ia'C^2,$$

$$D(\lambda_2) = a^2 + 2aa'C^2 + 2(a + a'C^2)a'C^2 - a^2 + 2(a')^2C^4$$
$$= 4a'C^2(a + a'C^2) \neq 0.$$

Therefore, in this case λ_2 are not roots of $D(\lambda)$.

Corollary 4.5.37 *(i) In the unstable case $a' > a/C^2$: both λ_2 are roots of $D(\lambda)$.*
(ii) If $a' < -a/C^2$ then neither of the λ_2 are roots of $D(\lambda)$.

Summarizing, we have proved the following result:

Theorem 4.5.38 *(i) If $a' \in (-\infty, a/(\sqrt{2}C^2))$, the only root of $D(\lambda)$ is $\lambda = 0$ with multiplicity 2.*

(ii) If $a' \in [a/\sqrt{2}C^2, a/C^2)$, there are four roots of $D(\lambda)$: zero (multiplicity two) and $\pm i|\lambda_2|$ (pure imaginary) with λ_2 as in (4.5.147).

(iii) If $a' = +a/C^2$, the only root of $D(\lambda)$ is $\lambda = 0$ multiplicity 4.

(iv) If $a' \in (a/C^2, +\infty)$, there are four roots of $D(\lambda)$: zero (multiplicity two) and $\pm|\lambda_2|$ with λ_2 as in (4.5.147). In particular there exists a positive root (linear instability).

Remark 4.5.39 Imagine reducing a' starting from a value greater than a/C^2. Initially there are two real roots, $\pm|\lambda_2|$, which approach zero as $a' \to a/C^2$ from above, giving rise to an increase of the multiplicity of the $\lambda = 0$ root to four when $a' = a/C^2$. As a' is reduced further below a/C^2 these two roots reappear as a pair of conjugate pure imaginary roots which move from zero to $\pm i\omega$ as a' goes from a/C^2 to $a/\sqrt{2}C^2$. When $a' = a/\sqrt{2}C^2$ these two roots touch the branch point (end of the continuous spectrum) and move onto an "unphysical branch" (on which the conditions (4.5.120) do not hold). As a' is reduced further these roots do not return to the physical branch and thus even when their magnitude becomes zero they do not coalesce with the physical $\lambda = 0$ root to increase its multiplicity and most importantly the spectrum is pure continuous apart from zero for $a' < a/C^2$.

5

Adiabatic Effective Dynamics of Solitons

In this chapter we present without proofs the results of [84] on *adiabatic effective dynamics* for the wave–particle system (1.5.1)–(1.5.2) in the case of *slowly varying external potential*. We also present a list of further generalizations and discuss the related *mass–energy equivalence* of Einstein.

5.1 Solitons in Slowly Varying External Potentials

In this section we describe the first result [84] on adiabatic effective dynamics. The solitons (2.1.3) are solutions to the system (1.5.1)–(1.5.2) with zero external potential $V(x) \equiv 0$. The asymptotic stability of the corresponding solitary manifold, proved in [127], means the *soliton-like asymptotics*

$$\psi(x, t) \approx \psi_{v(t)}(x - q(t)), \qquad t \in \mathbb{R} \tag{5.1.1}$$

for any solution with initial state sufficiently close to this manifold. On the other hand, solutions of this form may exist even for the system (1.5.1)–(1.5.2) with nonzero external potential if this potential is slowly varying:

$$|\nabla V(q)| \leq \varepsilon \ll 1. \tag{5.1.2}$$

In this case, the total momentum (2.1.2) is generally not conserved, but its slow evolution and the (fast) evolution of the parameter $q(t)$ in (5.1.1) can be described in terms of some finite-dimensional Hamiltonian dynamics.

Namely, denote by $P = P_v$ the total momentum of the soliton $S_{v,Q}$ in the notation (2.1.7). It is important that the map $\mathcal{P}: v \mapsto P_v$ is an isomorphism of the ball $|v| < 1$ on R^3. Therefore, we can consider Q, P as global coordinates on the solitary manifold $\mathcal{S}$. We define effective Hamiltonian functional

$$\mathcal{H}_{\text{eff}}(Q, P_v) \equiv \mathcal{H}(S_{v,Q}), \qquad Q, P_v \in \mathbb{R}^3, \tag{5.1.3}$$

where $\mathcal{H}$ is the total Hamiltonian (1.5.4). This functional allows the splitting $\mathcal{H}_{\text{eff}}(Q, \Pi) = E(\Pi) + V(Q)$ since the first integral in (1.5.4) does not depend on Q while the last integral vanishes on the solitons. Hence, the corresponding Hamiltonian equations read

$$\dot{Q}(t) = \nabla E(\Pi(t)), \qquad \dot{\Pi}(t) = -\nabla V(Q(t)). \tag{5.1.4}$$

The main result of [84] is the following theorem.

Theorem 5.1.1 *Let condition (5.1.2) hold, the initial state* $S_0 = (\psi_0, \pi_0, q_0, p_0) \in \mathcal{S}$ *is a soliton with total momentum* P_0, *and* $\psi(x,t), \pi(x,t), q(t), p(t)$ *of the system (1.5.1)–(1.5.2). Then the following "adiabatic asymptotics" hold*

$$|q(t) - Q(t)| \le C_0, \quad |P(t) - \Pi(t)| \le C_1 \varepsilon \quad \text{for} \quad |t| \le C \varepsilon^{-1}, \tag{5.1.5}$$

$$\sup_{t \in \mathbb{R}} \left[\|\nabla[\psi(q(t) + x, t) - \psi_{v(t)}(x)]\|_R + \|\pi(q(t) + x, t) - \pi_{v(t)}(x)\|_R \right] \le C \varepsilon, \tag{5.1.6}$$

where $P(t)$ *denotes total momentum (2.1.2),* $v(t) = \mathcal{P}^{-1}(\Pi(t))$, *and* $(Q(t), \Pi(t))$ *is the solution to the effective Hamiltonian equations (5.1.4) with initial conditions*

$$Q(0) = q(0), \qquad \Pi(0) = P(0).$$

Note that such relevance of effective dynamics (5.1.4) is due to the consistency of Hamiltonian structures:

(1) The effective Hamiltonian (5.1.3) is a restriction of the Hamiltonian functional (1.5.4) onto the soliton manifold $\mathcal{S}$.
(2) As shown in [84], the canonical form of the Hamiltonian system (5.1.4) is also a restriction onto $\mathcal{S}$ of canonical form of the system (1.5.1)–(1.5.2): formally

$$P \, dQ = \left[p \, dq + \int dx \, \pi(x) \, d\psi(x) \right] \Big|_{\mathcal{S}}.$$

Therefore, the total momentum P is canonically conjugate to the variable Q on the solitary manifold $\mathcal{S}$. This fact justifies definition (5.1.3) of the effective Hamiltonian as a function of the total momentum P_v, and not of the particle momentum p_v.

One of the important results of [84] is the following "effective dispersion relation":

$$E(\Pi) \sim \frac{\Pi^2}{2(1 + m_e)} + \text{const}, \qquad |\Pi| \ll 1. \qquad (5.1.7)$$

It means that nonrelativistic mass of a slow soliton increases due to an interaction with the field by the amount

$$m_e = -\frac{1}{3}\langle \rho, \Delta^{-1}\rho \rangle. \qquad (5.1.8)$$

This increment is proportional to the field energy of a soliton in rest

$$\mathcal{H}(\Delta^{-1}\rho, 0, 0, 0) = -\frac{1}{2}\langle \rho, \Delta^{-1}\rho \rangle,$$

which agrees with the Einstein mass–energy equivalence principle (see Section 5.2).

Remark 5.1.2 The relation (5.1.7) gives only a hint that m_e is an increment of the effective mass. The true *dynamical justification* for such an interpretation is given by the adiabatic asymptotics (5.1.5)–(5.1.6) which demonstrate the relevance of the effective dynamics (5.1.4).

Generalizations In [85], the asymptotics (5.1.5), (5.1.6) were extended to solitons of the Maxwell–Lorentz equations (1.6.1) with small external fields.

After the papers [84, 85] suitable adiabatic effective dynamics was obtained in [82, 83] for nonlinear Hartree and Schrödinger equations with slowly varying external potentials. Similar effective dynamics in the presence of small external fields later was constructed (i) in [81, 86, 87], for nonlinear systems of Einstein–Dirac, Chern–Simon–Schrödinger, and Klein–Gordon–Maxwell, and (ii) in [88], for Maxwell–Lorentz equations with rotating charge. Similar adiabatic effective dynamics was established in [80] for electrons in second-quantized Maxwell fields in the presence of a slowly varying external potential.

The results of numerical simulation [58] (see Chapter 6) confirm the adiabatic effective dynamics of solitons (5.1.6) for relativistic 1D nonlinear wave equations.

5.2 Mass–Energy Equivalence

In the case of the Maxwell–Lorentz equations [85] the increment of nonrelativistic mass also turns out to be proportional to the energy of the static soliton's own field.

Such equivalence of the self-energy of a particle with its mass was first discovered by M. Abraham in 1902: he obtained by direct calculation that electromagnetic self energy E_{own} of an electron at rest adds

$$m_e = \frac{4}{3} E_{\text{own}}/c^2$$

to its nonrelativistic mass (see [203, 204], and also [213, pp. 216–217]). It is easy to see that this self-energy is infinite for a point electron at the origin with a charge density $\delta(x)$, because in this case, the Coulomb electrostatic field $|E(x)| = C/|x|^2$, so the integral in (1.6.3) diverges around $x = 0$. This means that the field mass for a point electron is infinite, which contradicts experiments. That's why M. Abraham introduced the model of electrodynamics with an "extended electron" (1.6.1), whose self-energy is finite.

At the same time, M. Abraham conjectured that the *entire mass* of an electron is due to its own electromagnetic energy; that is, $m = m_e$: "*matter disappeared, only energy remains*"; see [210, pp. 63, 87, 88] (smile :)).

This conjecture was justified in 1905 by A. Einstein, who discovered the famous universal relation $E = m_0 c^2$, which follows from the Special Theory of Relativity [206]. The doubtful factor $\frac{4}{3}$ in the M. Abraham formula is due to the nonrelativistic character of the system (1.6.1). According to the modern view, about 80 percent of the electron mass is of electromagnetic origin [207].

6

Numerical Simulation of Solitons

In this chapter we describe the results of joint work with Arkady Vinnichenko (1945–2009) on numerical simulation of (i) global attraction to solitons (12) and (13) and (ii) adiabatic effective dynamics of solitons (5.1.6) for relativistic 1D nonlinear wave equations. Additional information can be found in [58].

6.1 Kinks of Relativistic Equations

First let us describe numerical simulations of solutions to relativistic 1D nonlinear wave equations with the polynomial nonlinearity

$$\ddot{\psi}(x,t) = \psi''(x,t) + F(\psi(x,t)), \qquad \text{where} \quad F(\psi) := -\psi^3 + \psi. \quad (6.1.1)$$

Since $F(\psi) = 0$ for $\psi = 0, \pm 1$, there are three stationary states: $S(x) \equiv 0, +1, -1$. This equation is formally equivalent to a Hamiltonian system (1.1.2) with the Hamiltonian

$$H(\psi,\pi) = \int \left[\frac{1}{2}|\pi(x)|^2 + \frac{1}{2}|\psi'(x)|^2 + U(\psi(x)) \right] dx \qquad (6.1.2)$$

where the potential is $U(\psi) = \frac{\psi^4}{4} - \frac{\psi^2}{2} + \frac{1}{4}$. This Hamiltonian is finite for functions $(\psi,\pi) \in \mathcal{E}$, where $\mathcal{E} = H_c^1 \oplus L^2$ (see (1.1.4)), for which the convergence

$$\psi(x) \to \pm 1, \qquad |x| \to \infty$$

is sufficiently fast.

The potential $U(\psi)$ has minima at $\psi = \pm 1$ and a maximum at $\psi = 0$. Accordingly, the two finite-energy solutions $\psi = \pm 1$ are stable, and the solution $\psi = 0$ with infinite-energy is unstable. Such potentials with two wells are called potentials of Ginzburg–Landau type.

In addition to the constant stationary solutions $S(x) \equiv 0, +1, -1$, there is also a nonconstant solution $S(x) = \tanh x/\sqrt{2}$, which is called a "kink." Its shifts and reflections $\pm S(\pm x - a)$ are also stationary solutions, as well as their Lorentz transformations

$$\pm S(\gamma(\pm x - a - vt)), \qquad \gamma = 1/\sqrt{1 - v^2}, \quad |v| < 1.$$

These are uniformly moving "traveling waves" (that is, solitons). The kink is strongly compressed when the velocity v is close to ± 1. This compression is known as the *Lorentz contraction*.

Numerical simulation Our numerical experiments show a decay of finite-energy solutions to a finite set of kinks and dispersive waves outside the kinks, which corresponds to the asymptotics of (13). The result of one of the experiments is shown in Figure 6.1: a finite-energy solution of the equation (6.1.1) decays to three kinks. Here the vertical line is the time axis, and the horizontal line is the space axis. The spatial scale redoubles at $t = 20$ and $t = 60$.

In Figure 6.1, the red color corresponds to the values $\psi > 1 + \varepsilon$, the blue color to the values $\psi < -1 - \varepsilon$, and the yellow color to the intermediate values $-1 + \varepsilon < \psi < 1 - \varepsilon$, where $\varepsilon > 0$ is sufficiently small. Thus, the yellow stripes (which are the lightest zones in the black-and-white version of the figure) represent the kinks, while the blue and red zones outside the yellow stripes are filled with dispersive waves.

For $t = 0$, the solution begins with a rather chaotic behavior, when there are no visible kinks. After 20 seconds, three separate kinks appear, which subsequently move almost uniformly.

The Lorentz contraction The left kink moves to the left at a low velocity $v_1 \approx 0.24$, the central kink is almost still, because its velocity $v_2 \approx 0.02$ is very small, and the right kink moves very rapidly with velocity $v_3 \approx 0.88$. The Lorentz spatial contraction $\sqrt{1 - v_k^2}$ is clearly visible in this picture: the central kink is the widest, the left is a bit narrower, and the right one is quite narrow.

The Einstein time delation The Einstein time delation is also very pronounced. Namely, all three kinks pulsate because of the presence of a nonzero eigenvalue in the equation linearized on the kink. Indeed, substituting $\psi(x,t) = S(x) + \varepsilon\varphi(x,t)$ in (6.1.1), we obtain in the first-order approximation the linearized equation

$$\ddot{\varphi}(x,t) = \varphi''(x,t) - 2\varphi(x,t) - V(x)\varphi(x,t), \qquad (6.1.3)$$

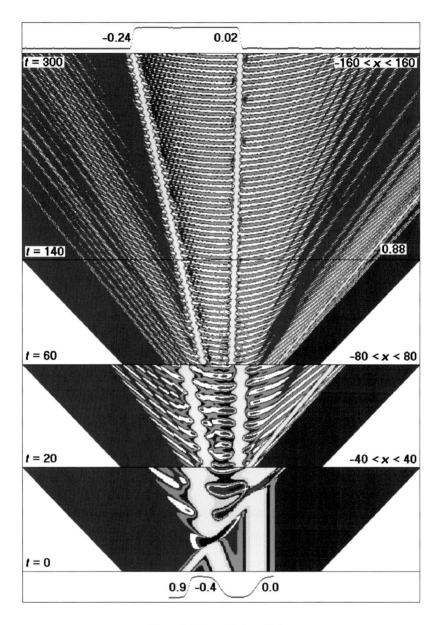

Figure 6.1 Decay to three kinks.

where the potential

$$V(x) = 3S^2(x) - 3 = -\frac{3}{\cosh^2 x/\sqrt{2}}$$

decays exponentially for large $|x|$. It is very fortunate that for this potential the spectrum of the corresponding *Schrödinger operator*

$$H := -\frac{d^2}{dx^2} + 2 + V(x)$$

is well known [59]. Namely, the operator H is non-negative, and its continuous spectrum is the interval $[2, \infty)$. It turns out that H also has a two-point discrete spectrum: the points $\lambda = 0$ and $\lambda = 3/2$. It is this nonzero eigenvalue which is responsible for the pulsations that we observe for the central slow kink, with frequency $\omega_2 \approx \sqrt{3/2}$ and period $T_2 \approx 2\pi/\sqrt{3/2} \approx 5$. On the other hand, for the fast kinks, the ripples are much slower, that is, the corresponding period is longer. This time delation agrees numerically with the Lorentz formulas, which confirms the relevance of these numerical simulation results.

Dispersive waves An analysis of dispersive waves provides additional confirmation. Namely, the space outside the kinks in Figure 6.1 is filled with dispersive waves whose values are very close to ± 1, with accuracy of 0.01. These waves satisfy with high accuracy the linear Klein–Gordon equation obtained by linearization of the Ginzburg–Landau equation (6.1.1) on the stationary solutions $\psi_\pm \equiv \pm 1$:

$$\ddot{\varphi}(x,t) = \varphi''(x,t) + 2\varphi(x,t).$$

The corresponding dispersion relation $\omega^2 = k^2 + 2$ determines the group velocities of high-frequency wave packets:

$$\omega'(k) = \frac{k}{\sqrt{k^2 + 2}} = \pm \frac{\sqrt{\omega^2 - 2}}{\omega}. \tag{6.1.4}$$

These wave packets are clearly visible in Figure 6.1 as straight lines whose propagation velocities converge to ± 1. This convergence is explained by the high-frequency limit $\omega'(k) \to \pm 1$ as $\omega \to \pm\infty$. For example, for dispersive waves emitted by the central kink, the frequencies $\omega = \pm n\omega_2 \to \pm\infty$ are generated by the polynomial nonlinearity in (6.1.1) in accordance with Figure 3.2.

Remark 6.1.1 These observations of dispersive waves agree with the radiative mechanism from Section 3.10.

The nonlinearity in (6.1.1) is chosen exactly because of the known spectrum of the linearized equation (6.1.3). In numerical experiments [58], more general nonlinearities of Ginzburg–Landau type have also been considered. The results were qualitatively the same: for "any" initial data of finite-energy, the solution decays for large times to a sum of kinks and of dispersive waves. Numerically, this is clearly visible, but rigorous justification remains an open problem.

6.2 Numerical Observation of Soliton Asymptotics

Besides the kinks the numerical experiments [58] also revealed soliton asymptotics of type (13) and adiabatic effective dynamics of the form (5.1.6) for complex solutions of the 1D relativistic nonlinear wave equations (2.2.4). Polynomial potentials of the form

$$U(\psi) = a|\psi|^{2m} - b|\psi|^{2n}, \qquad (6.2.1)$$

were considered with $a, b > 0$ and $m > n = 2, 3, \ldots$. Accordingly,

$$F(\psi) = 2am|\psi|^{2m-2}\psi - 2bn|\psi|^{2n-2}\psi. \qquad (6.2.2)$$

The parameters a, b, m, n were taken as follows,

N	a	m	b	n
1	1	3	0.61	2
2	10	4	2.1	2
3	10	6	8.75	5

Various "smooth" initial functions $\psi(x,0), \dot\psi(x,0)$ with supports on the interval $[-20, 20]$ were considered. The second-order difference scheme with $\Delta x \sim 0.01$ and $\Delta t \sim 0.001$ was employed. In all cases, asymptotics of type (13) were observed with the numbers 0, 1, 3, and 5 of solitons for $t > 100$.

6.3 Adiabatic Effective Dynamics of Relativistic Solitons

In the numerical experiments [58], the adiabatic effective dynamics of the form (5.1.6) was also observed for soliton-like solutions of type (5.1.1) of the 1D equations (2.2.4) with a slowly varying external potential (5.1.2):

$$\ddot\psi(x,t) = \psi''(x,t) - \psi(x,t) + F(\psi(x,t)) - V(x)\psi(x,t), \qquad x \in \mathbb{R}. \qquad (6.3.1)$$

This equation is formally equivalent to the Hamiltonian system (1.1.2) with the Hamiltonian

$$H_V(\psi, \pi) = \int \left[\frac{1}{2} |\pi(x)|^2 + \frac{1}{2} |\psi'(x)|^2 + U(\psi(x)) + \frac{1}{2} V(x) |\psi(x)|^2 \right] dx.$$
$$(6.3.2)$$

The soliton-like solutions are of the form (cf. (5.1.1))

$$\psi(x,t) \approx e^{i\Theta(t)} \phi_{\omega(t)}(\gamma_{v(t)}(x - q(t))). \qquad (6.3.3)$$

The numerical experiments [58] qualitatively confirm the Hamiltonian adiabatic effective Hamiltonian dynamics for the parameters Θ, ω, q, and v, but it has not yet been rigorously justified.

Figure 6.2 represents a solution to equation (6.3.1) with the potential (6.2.1), where $a = 10$, $m = 6$ and $b = 8.75$, $n = 5$. The potential is $V(x) = -0.2\cos(0.31x)$ and the initial conditions are

$$\psi(x,0) = \phi_{\omega_0}(\gamma_{v_0}(x - q_0)), \qquad \dot{\psi}(x,0) = 0, \qquad (6.3.4)$$

where $v_0 = 0$, $\omega_0 = 0.6$ and $q_0 = 5.0$. We note that the initial state does not belong to the solitary manifold. The effective width (half-amplitude) of the solitons is in the range [4.4, 5.6]. It is quite small when compared with the spatial period of the potential $2\pi/0.31 \sim 20$. The results of the numerical simulations are shown in Figure 6.2.

• The blue and green colors represent a dispersive wave with values $|\psi(x,t)| < 0.01$, while the red color represents the top of a soliton with values $|\psi(x,t)| \in [0.4, 0.8]$.

• The soliton trajectory (the "red snake" in the middle of Figure 6.2) corresponds to oscillations of a classical particle in the potential $V(x)$.

• For $0 < t < 140$, the solution is rather distant from the solitary manifold, and the radiation is rather intense.

• For $3020 < t < 3180$, the solution approaches the solitary manifold, and the radiation weakens. The amplitude of oscillations of the soliton is almost unchanged over a long time, confirming the Hamiltonian type of the effective dynamics.

• However, for $5260 < t < 5420$, the amplitude of the soliton oscillation is halved. This suggests that on a large time scale the deviation from Hamiltonian effective dynamics becomes essential. Consequently, the effective dynamics gives a good approximation only on an adiabatic time scale of type $t \sim \varepsilon^{-1}$.

• The deviation of the effective dynamics from being Hamiltonian is due to radiation, which plays the role of dissipation.

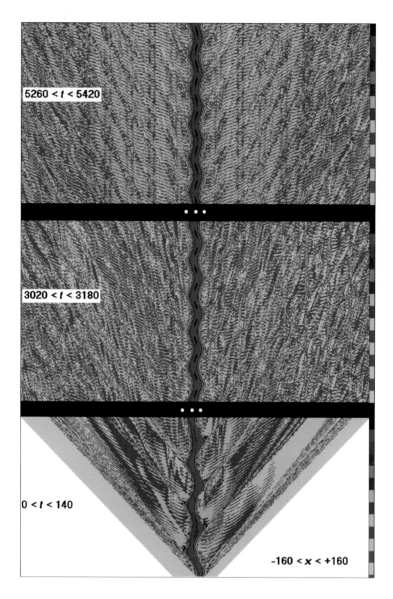

Figure 6.2 Adiabatic effective dynamics of relativistic solitons.

• The radiation is realized as dispersive waves, which carry energy to infinity. The dispersive waves combine into uniformly moving wave packets with a discrete set of group velocities, as in Figure 6.1. The magnitude of the solution is of order ~ 1 on the trajectory of the soliton, while the values of the dispersive waves are less than 0.01 for $t > 200$, so that their energy density does not exceed 0.0001. The amplitude of the dispersive waves decays at large times.

• In the limit as $t \rightarrow \pm\infty$, the soliton should converge to a static position corresponding to a local minimum of the potential $V(x)$. However, the numerical observation of this "ultimate stage" is hopeless, since the rate of the convergence decays with the decrease of the radiation.

7

Dispersive Decay

In this chapter we give (i) a brief survey of basic results on the dispersive decay and (ii) a new short and simplified proof of the fundamental results on the $L^1 \to L^\infty$ dispersive decay for the Schrödinger equation established in [184] by J.-L. Journé, A. Soffer, and C.D. Sogge.

7.1 The Schrödinger and Klein–Gordon Equations

7.1.1 Dispersive Decay in Weighted Sobolev Norms

A powerful systematic approach to dispersive decay in weighted Sobolev norms for the Schrödinger equation with potential was proposed by S. Agmon and A. Jensen and T. Kato [171, 183]. This theory was extended by many authors to wave, Klein–Gordon, and Dirac equations and to the corresponding discrete equations, see [10, 142, 143], [172]–[182], [184]–[197], [198, 199] and references therein.

7.1.2 Decay $L^1 - L^\infty$

$$\| P_c \psi(t) \|_{L^\infty(\mathbb{R}^n)} \leq C t^{-n/2} \| \psi(0) \|_{L^1(\mathbb{R}^n)}, \qquad t > 0 \qquad (7.1.1)$$

for solutions of the linear Schrödinger equation

$$i \dot{\psi}(x,t) = H \psi(x,t) := (-\Delta + V(x)) \psi(x,t), \qquad x \in \mathbb{R}^n \qquad (7.1.2)$$

with $n \geq 3$ was established for the first time by J.-L. Journé, A. Soffer and C.D. Sogge [184] provided that $\lambda = 0$ is neither an eigenvalue nor resonance for H. The potential $V(x)$ is sufficiently smooth and rapidly decays as $|x| \to \infty$. Here P_c is an orthogonal projection onto continuous spectral space of the operator H. This result was generalized later by many authors, see below.

In [199] a decay of type (7.1.1) and Strichartz estimates were established for 3D Schrödinger equations (7.1.2) with "rough" and time-dependent potentials $V = V(x,t)$ (in stationary case $V(x)$ belongs to both the Rollnik class and the Kato class). Similar estimates were received in [175] for 3D Schrödinger and wave equations with (stationary) Kato class potentials.

In [179] the 4D Schrödinger equations (7.1.2) are considered for the case when there is a resonance or an eigenvalue at zero energy. In particular, in the case of an eigenvalue at zero energy, there is a time-dependent operator F_t of rank 1, such that $\|F_t\|_{L^1 \to L^\infty} \leq 1/\log t$ for $t > 2$, and

$$\|e^{itH} P_c - F_t\|_{L^1 \to L^\infty} \leq Ct^{-1}, \qquad t > 2.$$

Similar dispersive estimates are proved also for solutions to 4D wave equation with a potential.

In [181, 182] the Schrödinger equation (7.1.2) is considered in $\mathbb{R}^n$ with $n \geq 5$ when there is an eigenvalue at the zero point of the spectrum. It is shown, in particular, that there is a time-dependent rank one operator F_t such that $\|F_t\|_{L^1 \to L^\infty} \leq C|t|^{2-n/2}$ for $|t| > 1$, and

$$\|e^{itH} P_c - F_t\|_{L^1 \to L^\infty} \leq C|t|^{1-n/2}, \qquad |t| > 1.$$

With a stronger decay of the potential, the evolution admits an operator-valued expansion

$$e^{itH} P_c(H) = |t|^{2-n/2} A_{-2} + |t|^{1-n/2} A_{-1} + |t|^{-n/2} A_0,$$

where A_{-2} and A_{-1} are finite rank operators $L^1(\mathbb{R}^n) \to L^\infty(\mathbb{R}^n)$, while A_0 maps weighted L^1 spaces to weighted L^∞ spaces. Main members A_{-2} and A_{-1} equal to zero under certain conditions of the orthogonality of the potential V to eigenfunction with zero energy. Under the same orthogonality conditions, the remainder term $|t|^{-n/2} A_0$ also maps $L^1(\mathbb{R}^n)$ to $L^\infty(\mathbb{R}^n)$, and therefore, the group $e^{itH} P_c(H)$ has the same dispersive decay as free evolution, despite its eigenvalue at zero.

7.1.3 Decay $L^p - L^q$ for the Klein–Gordon Equation

Such decay was first established in [198] for solutions of the free Klein–Gordon equation $\ddot{\psi} = \Delta\psi - \psi$ in $\mathbb{R}^n$ with initial state $\psi(0) = 0$:

$$\|\psi(t)\|_{L^q} \leq Ct^{-d} \|\dot{\psi}(0)\|_{L^p}, \qquad t > 1, \tag{7.1.3}$$

where $1 \leq p \leq 2$, $1/p + 1/q = 1$, and $d \geq 0$ is a piecewise-linear function of $(1/p, 1/q)$. The proofs use the Riesz interpolation theorem.

In [174], the estimates (7.1.3) are extended to solutions of the perturbed Klein–Gordon equation

$$\ddot{\psi} = \Delta\psi - \psi + V(x)\psi, \qquad x \in \mathbb{R}^n$$

with $\psi(0) = 0$. The authors show that (7.1.3) holds for $0 \leq 1/p - 1/2 \leq 1/(n+1)$. The smallest value of p and the fastest decay rate d occurs when $1/p = 1/2 + 1/(n+1)$, $d = (n-1)/(n+1)$. The result is proved under the assumption that the potential V is smooth and small in a suitable sense. For example, the result is true when $|V(x)| \leq c(1 + |x|^2)^{-\sigma}$, where $c > 0$ is sufficiently small. Here $\sigma > 2$ for $n = 3$, $\sigma > n/2$ for odd $n \geq 5$, and $\sigma > (2n^2 + 3n + 3)/4(n+1)$ for even $n \geq 4$. The results also apply to the case when $\psi(0) \neq 0$.

7.1.4 Decay $L^p - L^q$ for the Schrödinger Equation

The seminal article [184] concerns $L^p - L^q$ decay of solutions to the Schrödinger equation (7.1.2). It is assumed that $(1 + |x|^2)^\alpha V(x)$ is a multiplier in the Sobolev spaces H^η for some $\eta > 0$ and $\alpha > n + 4$, and the Fourier transform of V belongs to $L^1(\mathbb{R}^n)$. Under these conditions, the main result of [184] is the following theorem: if $\lambda = 0$ is neither an eigenvalue nor a resonance for H, then

$$\|P_c\psi(t)\|_{L^q} \leq Ct^{-n(1/p-1/2)}\|\psi(0)\|_{L^p}, \qquad t > 1, \qquad (7.1.4)$$

where $1 \leq p \leq 2$ and $1/p + 1/q = 1$. Proofs are based on $L^1 - L^\infty$ decay (7.1.1) and the Riesz interpolation theorem.

In [202] estimates (7.1.4) are proved for the Schrödinger equation (7.1.2) under suitable conditions on the decay of $V(x)$ (i) with $1 \leq p \leq 2$ if $\lambda = 0$ is neither an eigenvalue nor a resonance for H and (ii) with all $3/2 < p \leq 2$ otherwise.

7.2 Decay $L^1 - L^\infty$ for 3D Schrödinger Equations

In this section we give a new short and simplified proof of the $L^1 - L^\infty$ dispersive decay (7.1.1) for the Schrödinger equation (7.1.2), first established in [184] for $n \geq 3$. We restrict ourselves by the case $n = 3$,

$$i\dot{\psi}(x,t) = -\Delta\psi(x,t) + V(x)\psi(x,t), \qquad x \in \mathbb{R}^3. \tag{7.2.1}$$

Another approach to the proof of this decay in the cases $n = 1$ and $n = 3$ was suggested by M. Goldberg and W. Schlag [180].

Our approach [195] considerably simplifies the arguments of [184] and [180]. We suppose that the potential $V(x)$ is a continuous real function, and

$$|V(x)| \leq C\langle x\rangle^{-\beta}, \qquad \langle x\rangle = (1 + |x|^2)^{1/2}, \qquad x \in \mathbb{R}^3 \tag{7.2.2}$$

for some $\beta > 3$. As in [184] and [180] we consider the "regular case" when the point 0 is neither eigenvalue nor resonance for the Schrödinger operator $H = -\Delta + V(x)$. Equivalently, the truncated resolvent of the operator $\mathcal{H}$ is bounded at the edge point of the continuous spectrum.

Theorem 7.2.1 *Let condition (7.2.2) with $\beta > 3$ hold. Then in the regular case,*

$$\|e^{itH}P_c(H)\|_{L^1 \to L^\infty} \leq C|t|^{-\frac{3}{2}}, \quad |t| \geq 1, \tag{7.2.3}$$

where $P_c(H)$ is the Riesz projection onto the continuous spectrum of H.

This theorem immediately implies the decay in weighted norms

$$\|\psi\|_{L^p_\sigma} = \|\langle x\rangle^\sigma \psi\|_{L^p}, \qquad \sigma \in \mathbb{R}.$$

Corollary 7.2.2 *Let (7.2.2) hold and $\sigma > 3/2$. Then in the regular case,*

$$\|e^{itH}P_c(H)\|_{L^2_\sigma \to L^2_{-\sigma}} \leq C(1 + |t|)^{-3/2}, \quad t \in \mathbb{R}. \tag{7.2.4}$$

Indeed, for any bounded operator $K: L^1 \to L^\infty$ and any $f \in L^2_\sigma$ with $\sigma > 3/2$, one has

$$\|Kf\|_{L^2_{-\sigma}} \leq C\|Kf\|_{L^\infty} \leq C\|K\|_{L^1 \to L^\infty}\|f\|_{L^1} \leq C_1\|K\|_{L^1 \to L^\infty}\|f\|_{L^2_\sigma}.$$

Remark 7.2.3 For $\sigma > 5/2$ the dispersive decay of type (7.2.4) for the 3D Schrödinger equation was established first by A. Jensen and T. Kato [183].

Our proofs follow the general strategy of [171, 180, 183] which relies on the spectral Fourier representation

$$e^{iHt} P_c(H) = \frac{1}{2\pi i} \int_0^\infty e^{-i\omega t} \Big[R(\omega + i0) - R(\omega - i0) \Big] d\omega, \qquad (7.2.5)$$

where $R(\omega) = (H - \omega)^{-1}$ is the resolvent of the Schrödinger operator H.

We verify the decay (7.2.3) of the integral (7.2.5) developing a streamlined version of the approach [180]. First note that this integral generally does not converge in the operator norm $L_\sigma^2 \to L_{-\sigma}^2$ due to the slow decay of the resolvent in this norm like $\sim \omega^{-1/2}$ by the results of S. Agmon and A. Jensen and T. Kato [171, 183]. On the other hand, this integral converges in the sense of distributions, that is the integrals over intervals $[0, L]$ are tempered distributions $W_L(t)$ which converge as $L \to \infty$. Thus, (7.2.3) will follow from uniform in $L \geq 1$ estimates

$$\|W_L(t)\|_{L^1 \to L^\infty} \leq C|t|^{-\frac{3}{2}}, \quad |t| \geq 1. \qquad (7.2.6)$$

Note that for the free Schrödinger operator, the decay holds since its integral kernel is bounded for $|t| \geq 1$ and decays uniformly in space:

$$\|e^{-i\Delta t}\|_{L^1 \to L^\infty} = \sup_{x, y \in \mathbb{R}^3} \left| \frac{e^{i|x-y|^2/4t}}{(4\pi i t)^{3/2}} \right| \leq C|t|^{-3/2}, \quad |t| \geq 1. \qquad (7.2.7)$$

7.2.1 Properties of the Resolvent

Here we collect the properties of the resolvent $R(\omega) = (H - \omega)^{-1}$ obtained in [171, 183] (see also [188] where the full proofs of these properties can be found). We suppose that the condition (7.2.2) holds with some $\beta > 1$. Then

R1. $R(\omega) : L^2 \to L^2$ is a meromorphic function of $\omega \in \mathbb{C} \setminus [0, \infty)$; the poles of $R(\omega)$ are located at a finite set of eigenvalues $\omega_j < 0$.

R2. For $\omega > 0$ and $\sigma > 1/2$, there exist the limits $R(\omega \pm i0)$ such that

$$\|R(\omega \pm i\varepsilon) - R(\omega \pm i0)\|_{L_\sigma^2 \to L_{-\sigma}^2} \to 0, \quad \varepsilon \to 0+.$$

R3. Let $\beta > 3$. Then for $\sigma > 1/2 + k$,

$$\|R^{(k)}(\omega)\|_{L_\sigma^2 \to L_{-\sigma}^2} = \mathcal{O}(|\omega|^{-\frac{1+k}{2}}), \quad |\omega| \to \infty, \quad \omega \in \mathbb{C} \setminus [0, \infty), \quad k = 0, 1, 2. \qquad (7.2.8)$$

R4. Let $\beta > 2$. Then in the regular case, $R^{\pm}(\omega) := R(\omega \pm i0)$ are continuous operator functions of $\omega \geq 0$ with the values in $B(L_\sigma^2, L_{-\sigma}^2)$ for any $\sigma_1, \sigma_2 > 1/2$ with $\sigma_1 + \sigma_2 > 2$.

R5. Let $\beta > 3$. Then in the regular case,

$$\|R^\pm(\omega)\|_{L^2_{\sigma_1} \to L^2_{-\sigma_2}} = \mathcal{O}(1), \quad \omega \to 0, \quad \sigma_1, \sigma_2 > 1/2, \quad \sigma_1 + \sigma_2 > 2,$$
$$(7.2.9)$$

$$\|\partial_\omega^k R^\pm(\omega)\|_{L^2_\sigma \to L^2_{-\sigma}} = \mathcal{O}(|\omega|^{\frac{1}{2}-k}), \quad \omega \to 0, \quad \sigma > 1/2 + k, \quad k = 1, 2.$$
$$(7.2.10)$$

In particular, all these properties hold for the free resolvent $R_0(\omega) = (-\Delta - \omega)^{-1}$.

The integral kernels of $R_0^\pm(\lambda^2)$ have an explicit representation

$$R_0^\pm(\lambda^2, x, y) = \frac{e^{\pm i\lambda|x-y|}}{4\pi |x - y|}. \qquad (7.2.11)$$

The asymptotics (7.2.9)–(7.2.10) imply

Lemma 7.2.4 *Let (7.2.2) hold with $\beta > 3$. Then in the regular case*

$$\|\partial_\lambda^k R^\pm(\lambda^2)\|_{L^2_{\sigma_1} \to L^2_{-\sigma_2}} \leq C(1 + \lambda)^{-1}, \quad \lambda \geq 0, \qquad (7.2.12)$$

where $\sigma_1, \sigma_2 > 1/2$, $\sigma_1 + \sigma_2 > 2$ for $k = 0$, and $\sigma_1, \sigma_2 > 1/2 + k$ for $k = 1, 2$.

Proof First, note that

$$\|\partial_\lambda^k R^\pm(\lambda^2)\|_{L^2_{\sigma_1} \to L^2_{-\sigma_2}} = \mathcal{O}(1), \quad \lambda \to 0, \qquad (7.2.13)$$

where $\sigma_1, \sigma_2 > 1/2$, $\sigma_1 + \sigma_2 > 2$ for $k = 0$, and $\sigma_1, \sigma_2 > 1/2 + k$ for $k = 1, 2$. Indeed, asymptotics (7.2.13) with $k = 0$ follows from (7.2.9). Next we apply the formulas (see for example [188, Formulas (17.9), (17.11)])

$$R' = R_0' + RVR_0' + R_0'VR + RVR_0'VR,$$

$$R'' = R_0'' + RVR_0'' + R_0''VR + RVR_0''VR + 2R'VR_0' + 2R'VR_0'VR.$$
$$(7.2.14)$$

These formulas and (7.2.9), (7.2.10) imply asymptotics (7.2.13) with $k = 1, 2$. Similarly, asymptotics (7.2.8) together with formulas (7.2.14) imply

$$\left\| \frac{\partial^k}{\partial \lambda^k} R^\pm(\lambda^2) \right\|_{L^2_{\sigma_1} \to L^2_{-\sigma_2}}$$

$$= \mathcal{O}(\lambda^{-1}), \quad \lambda \to \infty, \quad \sigma_1, \sigma_2 > 1/2 + k, \quad k = 0, 1, 2. \qquad (7.2.15)$$

$\square$

7.2.2 The Born Series

The identity $R(\lambda^2) - R_0(\lambda^2) = -R(\lambda^2)V R_0(\lambda^2)$ implies that

$$R(\lambda^2) = (1 - R(\lambda^2)V)R_0(\lambda^2).$$

The iteration yields the finite Born series [180]

$$R^\pm = \sum_{k=0}^{N}(R_0^\pm V)^k R_0^\pm + (R_0^\pm V)^N R^\pm V R_0^\pm, \quad N \geq 0. \qquad (7.2.16)$$

Substituting the expansion with $N = 2$ into spectral representation (7.2.5), we obtain

$$e^{itH}P_c(H) = \sum_{j=0}^{2} S_j(t) + Z(t),$$

where $S_0(t) = e^{-i\Delta t}$, and

$$S_j(t) = \frac{1}{2\pi i} \int_0^\infty e^{-i\omega t}((R_0^+(\omega)V)^j R_0^+(\omega)$$

$$- (R_0^-(\omega)V)^j R_0^-(\omega))d\omega, \quad j = 1, 2, \qquad (7.2.17)$$

$$Z(t) = \frac{1}{2\pi i} \int_0^\infty e^{-i\omega t}((R_0^+(\omega)V)^2 R^+(\omega)V R_0^+(\omega)$$

$$- (R_0^-(\omega)V)^2 R^-(\omega)V R_0^-(\omega))d\omega. \qquad (7.2.18)$$

By (7.2.7),

$$\|S_0(t)\|_{L^1 \to L^\infty} \leq C|t|^{-\frac{3}{2}}, \quad |t| \geq 1.$$

It remains to prove similar decay for $S_j(t)$, $j = 1, 2$ and for $Z(t)$.

7.2.3 The Decay of $S_1(t)$ and $S_2(t)$

Lemma 7.2.5 *Let condition (7.2.2) hold with $\beta > 3$. Then for $j = 1, 2$*

$$\|S_j(t)\|_{L^1 \to L^\infty} \leq C|t|^{-\frac{3}{2}}, \quad |t| \geq 1. \qquad (7.2.19)$$

Proof Similarly to (7.2.6) it suffices to prove the decay

$$\sup_{L \geq 1} \|S_j(t, L)\|_{L^1 \to L^\infty} \leq C|t|^{-\frac{3}{2}}, \quad |t| \geq 1, \quad j = 1, 2, \qquad (7.2.20)$$

where

$$S_j(t,L) = \frac{1}{\pi i} \int_0^\infty e^{-i\lambda^2 t} \chi(\lambda/L)[(R_0^+(\lambda^2)V)^j R_0^+(\lambda^2)$$

$$- (R_0^-(\lambda^2)V)^j R_0^-(\lambda^2)]\lambda d\lambda \qquad (7.2.21)$$

and $\chi(\lambda) \in C_0^\infty(\mathbb{R})$ with $\chi(\lambda) = 1$ for $|\lambda| \le 1$.

The representation (7.2.11) implies

$$\sup_{x,y \in \mathbb{R}^3} |R_0^+(\lambda^2,x,y) - R_0^-(\lambda^2,x,y)| \le C\lambda, \qquad \lambda \ge 0. \qquad (7.2.22)$$

Hence,

$$\sup_{x,y \in R^3} |[R_0^+ V R_0^+ - R_0^- V R_0^-](\lambda^2,x,y)|$$

$$\le \sup_{x,y \in R^3} |[(R_0^+ - R_0^-)V R_0^+](\lambda^2,x,y)| + \sup_{x,y \in R^3} |[R_0^- V(R_0^+ - R_0^-)](\lambda^2,x,y)|$$

$$\le C\lambda \sup_{x,y \in R^3} \int \left(\frac{|V(z)|}{|x-z|} + \frac{|V(z)|}{|y-z|} \right) dz \le C_1 \lambda, \qquad \lambda \ge 0. \qquad (7.2.23)$$

Similarly,

$$\sup_{x,y \in R^3} |[(R_0^+ V)^2 R_0^+ - (R_0^- V)^2 R_0^-](\lambda^2,x,y)|$$

$$\le C\lambda \sup_{x,y \in R^3} \iint \left(\frac{|V(x_1)V(y_1)|}{|x_1-y_1||y_1-y|} + \frac{|V(x_1)V(y_1)|}{|x-x_1||x_1-y_1|} \right.$$

$$\left. + \frac{|V(x_1)V(y_1)|}{|x-x_1||y-y_1|} \right) dx_1 dy_1 \le C\lambda.$$

Therefore, we can integrate by parts in (7.2.21):

$$S_j(t,L) = \frac{1}{2\pi t} \int_0^\infty e^{-i\lambda^2 t} \partial_\lambda \big(\chi(\lambda/L)[(R_0^-(\lambda^2)V)^j R_0^-(\lambda^2)$$

$$- (R_0^+(\lambda^2)V)^j R_0^+(\lambda^2)](\lambda^2) \big) d\lambda$$

$$= \frac{1}{2\pi t}(T_j^-(t,L) - T_j^+(t,L)). \qquad (7.2.24)$$

It remains to prove that

$$\sup_{L \ge 1} \sup_{x,y} |T_j^\pm(t,L,x,y)| \le C|t|^{-1/2}, \qquad |t| \ge 1, \quad j = 1,2. \qquad (7.2.25)$$

We have

$$|T_1^\pm(t,L,x,y)| \le \frac{1}{L}\left|\int \frac{V(z)}{|z-y||z-x|}\left(\int_0^\infty e^{-i\psi_1^\pm(\lambda)t}\chi'(\lambda/L)d\lambda\right)dz\right|$$

$$+\left|\int_{\mathbb{R}^3}\left(\frac{V(z)}{|z-y|}+\frac{V(z)}{|z-x|}\right)\left(\int_0^\infty e^{-i\psi_1^\pm(\lambda)t}\chi(\lambda/L)d\lambda\right)dz\right|,$$

$$|T_2^\pm(t,L,x,y)|$$

$$\le \frac{1}{L}\left|\iint \frac{|V(x_1)||V(y_1)|}{|x-x_1||x_1-y_1||y_1-y|}\left(\int_0^\infty e^{-i\psi_2^\pm(\lambda)t}\chi'(\lambda/L)d\lambda\right)dx_1dy_1\right|$$

$$+\left|\iint\left(\frac{V(x_1)V(y_1)}{|x_1-y_1||y_1-y|}+\frac{V(x_1)V(y_1)}{|x-x_1||x_1-y_1|}+\frac{V(x_1)V(y_1)}{|x-x_1||y-y_1|}\right)\right.$$

$$\left.\times\left(\int_0^\infty e^{-i\psi_2^\pm(\lambda)t}\chi(\lambda/L)d\lambda\right)dx_1dy_1\right|,$$

where

$$\psi_1^\pm(\lambda) = \lambda^2 \mp \lambda(|x-z|+|z-y|)/t,$$

$$\psi_2^\pm(\lambda) = \lambda^2 \mp \lambda(|x-x_1|+|x_1-y|+|y_1-y|)/t$$

with

$$\partial_\lambda^2\psi_j^\pm(\lambda) = 2, \quad \lambda > 0, \quad j = 1,2.$$

Then (7.2.25) follows by the Van der Corput lemma (see [200, Chapter VIII, Proposition II and Corollary]). □

7.2.4 The Decay of $Z(t)$

Now the proof of Theorem 7.2.1 is reduced to the proof of the following proposition.

Proposition 7.2.6 *Let the conditions of Theorem 7.2.1 hold. Then*

$$\|Z(t)\|_{L^1\to L^\infty} \le C|t|^{-\frac{3}{2}}, \quad |t| \ge 1. \tag{7.2.26}$$

Proof Using the definition (7.2.28), we represent $Z(t)$ as

$$Z(t) = \frac{1}{\pi i} \int\limits_0^\infty e^{-i\lambda^2 t} \big(\Lambda^+(\lambda) - \Lambda^-(\lambda)\big) \lambda d\lambda. \qquad (7.2.27)$$

Here

$$\Lambda^\pm(\lambda) = (R_0^\pm(\lambda^2)V)^2 R^\pm(\lambda^2) V R_0^\pm(\lambda^2) = R_0^\pm(\lambda^2) V \Pi^\pm(\lambda) V R_0^\pm(\lambda^2), \qquad (7.2.28)$$

where we denote $\Pi^\pm(\lambda) = R_0^\pm(\lambda^2) V R^\pm(\lambda^2)$.

First we prove some properties of $\Lambda^\pm(\lambda)$. We denote by $a+$ any number $a + \varepsilon$ with an arbitrary small, but fixed $\varepsilon > 0$.

Lemma 7.2.7 *Let (7.2.2) hold with some $\beta > 3$. Then in the regular case,*

$$\|\Lambda^+(\lambda) - \Lambda^-(\lambda)\|_{L^1 \to L^\infty} \to 0, \quad \lambda \to 0. \qquad (7.2.29)$$

Proof For any $f, g \in L^1$, we obtain

$$|\langle f, (\Lambda^+(\lambda) - \Lambda^-(\lambda))g\rangle|$$

$$\leq |\langle V(R_0^-(\lambda^2) - R_0^+(\lambda^2))f, \Pi^+(\lambda)V R_0^+(\lambda^2)g\rangle|$$

$$\quad + |\langle V R_0^+(\lambda^2)f, (\Pi^+(\lambda) - \Pi^-(\lambda))V R_0^+(\lambda^2)g\rangle|$$

$$\quad + |\langle V R_0^+(\lambda^2)f, \Pi^-(\lambda)V(R_0^+(\lambda^2) - R_0^-(\lambda^2))g\rangle|$$

$$\leq \|V(R_0^-(\lambda^2) - R_0^+(\lambda^2))f\|_{L_{1+}^2} \|\Pi^+(\lambda)\|_{L_{1+}^2 \to L_{-1-}^2} \|V R_0^+(\lambda^2)g\|_{L_{1+}^2}$$

$$\quad + \|V R_0^+(\lambda^2)f\|_{L_{1+}^2} \|(\Pi^+(\lambda) - \Pi^-(\lambda))\|_{L_{1+}^2 \to L_{-1-}^2} \|V R_0^+(\lambda^2)g\|_{L_{1+}^2}$$

$$\quad + \|V R_0^+(\lambda^2)f\|_{L_{1+}^2} \|\Pi^-(\lambda)\|_{L_{1+}^2 \to L_{-1-}^2} \|V(R_0^+(\lambda^2) - R_0^-(\lambda^2))g\|_{L_{1+}^2}$$

since $(R_0^\pm)^* = R_0^\mp$. For any $0 \leq \sigma \leq \beta - 1/2$, we have

$$\|V R_0^\pm(\lambda^2)f\|_{L_\sigma^2}^2 \leq \int V^2(x)\langle x\rangle^{2\sigma} \left|\int R_0^\pm(\lambda^2, x, y)f(y)dy\right|^2 dx$$

$$\leq C \iint |f(y_1)||f(y_2)| \left(\int \frac{\langle x\rangle^{2\sigma - 2\beta}}{|x - y_1||x - y_2|}dx\right) dy_1 dy_2$$

$$\leq C_1 \|f\|_{L^1}^2. \qquad (7.2.30)$$

Further, for any $0 \le \sigma \le \beta - 3/2$,

$$\|V(R_0^\pm(\lambda^2) - R_0^\mp(\lambda^2))f\|_{L_\sigma^2}^2 \le C\lambda^2 \iiint \langle x \rangle^{2\sigma-2\beta}|f(y_1)||f(y_2)|dxdy_1dy_2$$

$$\le C_1\lambda^2\|f\|_{L^1}^2 \tag{7.2.31}$$

by (7.2.22). Finally,

$$\|\Pi^+(\lambda) - \Pi^-(\lambda)\|_{L_{1+}^2 \to L_{-1-}^2} \to 0, \quad \lambda \to 0. \tag{7.2.32}$$

Indeed, property **R4** implies

$$\|R_0^+(\lambda^2) - R_0^-(\lambda^2)\|_{L_{1+}^2 \to L_{-1-}^2} \to 0,$$

$$\|R^+(\lambda^2) - R^-(\lambda^2)\|_{L_{1+}^2 \to L_{-1-}^2} \to 0, \quad \lambda \to 0,$$

while $V : L_{-1-} \to L_{1+}$ is bounded for $\beta > 2$. $\qquad\square$

Lemma 7.2.8 *Let (7.2.2) hold with some $\beta > 3$. Then in the regular case,*

$$\|\partial_\lambda^k \Lambda^\pm(\lambda)\|_{L^1 \to L^\infty} \le C(1+\lambda)^{-2}, \quad \lambda \ge 0, \ k = 0, 1. \tag{7.2.33}$$

Proof We omit the signs $\pm$ not to overburden the exposition. For example, $R_0(\lambda^2)$ means $R_0^+(\lambda^2)$ or $R_0^-(\lambda^2)$, $\Pi(\lambda) = \Pi^\pm(\lambda)$, etc. First, we show that

$$\|\Pi(\lambda)\|_{L_{\sigma_1}^2 \to L_{-\sigma_2}^2} + \|\partial_\lambda \Pi(\lambda)\|_{L_{\frac{3}{2}+}^2 \to L_{-\frac{3}{2}-}^2} + \|\partial_\lambda^2 \Pi_N^\pm(\lambda)\|_{L_{\frac{5}{2}+}^2 \to L_{-\frac{5}{2}-}^2}$$

$$\le C(1+\lambda)^{-2}, \lambda \ge 0, \tag{7.2.34}$$

where $\sigma_1, \sigma_2 > 1/2$, $\sigma_1 + \sigma_2 > 2$. Indeed, (7.2.12)–(7.2.15) imply

$$\|\Pi(\lambda)\|_{L_{\sigma_1}^2 \to L_{-\sigma_2}^2} \le \|R_0(\lambda^2)\|_{L_{3/2+}^2 \to L_{-\sigma_2}^2}\|V\|_{L_{-3/2-}^2 \to L_{3/2+}^2}\|R(\lambda^2)\|_{L_{\sigma_1}^2 \to L_{-3/2-}^2}$$

$$\le C(1+\lambda)^{-2}.$$

Further, (7.2.12)–(7.2.15) imply

$$\|\partial_\lambda \Pi(\lambda)\|_{L_{\frac{3}{2}+}^2 \to L_{-\frac{3}{2}-}^2}$$

$$\le \|\partial_\lambda R_0(\lambda^2)\|_{L_{\frac{3}{2}+}^2 \to L_{-\frac{3}{2}-}^2}\|V\|_{L_{-\frac{3}{2}-}^2 \to L_{\frac{3}{2}+}^2}\|R(\lambda^2)\|_{L_{\frac{3}{2}+}^2 \to L_{-\frac{3}{2}-}^2}$$

$$+ \|R_0(\lambda^2)\|_{L_{\frac{3}{2}+}^2 \to L_{-\frac{3}{2}-}^2}\|V\|_{L_{-\frac{3}{2}-}^2 \to L_{\frac{3}{2}+}^2}\|\partial_\lambda R(\lambda^2)\|_{L_{\frac{3}{2}+}^2 \to L_{-\frac{3}{2}-}^2}$$

$$\le C(1+\lambda)^{-2},$$

$$\|\partial_\lambda^2 \Pi(\lambda)\|_{L^2_{\frac{5}{2}+} \to L^2_{-\frac{5}{2}-}}$$

$$\leq \|\partial_\lambda^2 R_0(\lambda^2)\|_{L^2_{\frac{5}{2}+} \to L^2_{-\frac{5}{2}-}} \|V\|_{L^2_{-\frac{1}{2}-} \to L^2_{\frac{5}{2}+}} \|R(\lambda^2)\|_{L^2_{\frac{5}{2}+} \to L^2_{-\frac{1}{2}-}}$$

$$\|\partial_\lambda R_0(\lambda^2)\|_{L^2_{\frac{3}{2}+} \to L^2_{-\frac{5}{2}-}} \|V\|_{L^2_{-\frac{3}{2}-} \to L^2_{\frac{3}{2}+}} \|\partial_\lambda R(\lambda^2)\|_{L^2_{\frac{5}{2}+} \to L^2_{-\frac{3}{2}-}}$$

$$+ \|R_0(\lambda^2)\|_{L^2_{\frac{1}{2}+} \to L^2_{-\frac{5}{2}-}} \|V\|_{L^2_{-\frac{5}{2}-} \to L^2_{\frac{1}{2}+}} \|\partial_\lambda^2 R(\lambda^2)\|_{L^2_{\frac{5}{2}+} \to L^2_{-\frac{5}{2}-}}$$

$$\leq C(1 + \lambda)^{-2}.$$

Now, (7.2.30) and (7.2.34) imply for any $f, g \in L^1$ and $k = 0, 1$

$$|\langle f, R_0(\lambda^2) V \partial_\lambda^k \Pi(\lambda) V R_0(\lambda^2) g \rangle|$$

$$= |\langle V R_0^*(\lambda^2) f, \partial_\lambda^k \Pi(\lambda) V R_0(\lambda^2) g \rangle|$$

$$\leq \|V R_0^*(\lambda^2) f\|_{L^2_{\frac{3}{2}+}} \|\partial_\lambda^k \Pi(\lambda)\|_{L^2_{\frac{3}{2}+} \to L^2_{-\frac{3}{2}-}} \|V R_0(\lambda^2) g\|_{L^2_{\frac{3}{2}+}}$$

$$\leq C(1 + \lambda)^{-2} \|f\|_{L^1} \|g\|_{L^1}. \tag{7.2.35}$$

Further, for any $0 \leq \sigma \leq \beta - 3/2$, we obtain

$$\|V \partial_\lambda R_0(\lambda^2) f\|_{L^2_\sigma}^2 \leq C \int \langle x \rangle^{2\sigma - 2\beta} \left(\int |f(y)| dy \right)^2 dx \leq C_1 \|f\|_{L^1}^2. \tag{7.2.36}$$

Hence, (7.2.30) and (7.2.34) imply for any $f, g \in L^1$ and $\lambda \geq 0$,

$$|\langle f, \partial_\lambda R_0(\lambda^2) V \Pi(\lambda) V R_0(\lambda^2) g \rangle|$$

$$= |\langle V \partial_\lambda R_0^*(\lambda^2) f, \Pi(\lambda) V R_0(\lambda^2) g \rangle|$$

$$\leq \|V \partial_\lambda R_0^*(\lambda^2) f\|_{L^2_{1+}} \|\Pi(\lambda)\|_{L^2_{1+} \to L^2_{-1-}} \|V R_0(\lambda^2) g\|_{L^2_{1+}}$$

$$\leq C \|f\|_{L^1} \|g\|_{L^1} (1 + \lambda)^{-2}. \tag{7.2.37}$$

Similarly,

$$|\langle f, R_0(\lambda^2) V \Pi(\lambda) V \partial_\lambda R_0(\lambda^2) g \rangle| \leq C \|f\|_{L^1} \|g\|_{L^1} (1 + \lambda)^{-2}, \quad \lambda \geq 0.$$

Then (7.2.33) follows by definition (7.2.28) of Λ. $\qquad\square$

Due to Lemma 7.2.8, the integrand in (7.2.27) is a differentiable operator function of $\lambda \geq 0$ with values in the space of bounded operators mapping L^1 into L^∞. Moreover, due to Lemmas 7.2.7 and 7.2.8, we can integrate by parts,

$$Z(t) = \frac{1}{2\pi t} \int_0^\infty e^{-i\lambda^2 t} \partial_\lambda \big(\Lambda^-(\lambda) - \Lambda^+(\lambda)\big) d\lambda = \frac{1}{2\pi t}(Q^-(t) - Q^+(t)).$$

Here

$$Q^\pm(t,x,y)$$
$$= \int_0^\infty \Big(e^{-i\varphi_1 t} K_1^\pm(\lambda,x,y) + e^{-i\varphi_2^\pm t} K_2^\pm(\lambda,x,y) e^{-i\varphi_3^\pm t} K_3^\pm(\lambda,x,y)\Big) d\lambda,$$

where we denote

$$\varphi_1(\lambda) = -\lambda^2, \quad \varphi_2^\pm(\lambda) = -\lambda^2 \mp \lambda|x|/t, \quad \varphi_3^\pm(\lambda) = -\lambda^2 \mp \lambda|y|/t,$$

$$K_1^\pm(\lambda) = R_0^\pm(\lambda^2) V \partial_\lambda \Pi_{N-1}^\pm(\lambda) V R_0^\pm(\lambda^2),$$

$$K_2^\pm(\lambda) = (G^\mp(\lambda))^* V \Pi_N^\pm(\lambda) V R_0^\pm(\lambda^2), \quad K_3^\pm(\lambda) = R_0^\pm(\lambda^2) V \Pi_N^\pm(\lambda) V G^\pm(\lambda),$$

and $G^\pm(\lambda)$ is the operator with the kernel

$$G^\pm(\lambda,x,y) = \mp \frac{e^{\pm i\lambda(|x-y|-|y|)}}{4\pi i}, \quad \lambda \geq 0. \tag{7.2.38}$$

It remains to prove that

$$\sup_{x,y} |Q^\pm(t,x,y)| \leq C|t|^{-\frac{1}{2}}, \quad |t| \geq 1. \tag{7.2.39}$$

To this end, we estimate the functions $K_j^\pm$, $j = 1, 2, 3$.

Lemma 7.2.9 *Let (7.2.2) hold with some $\beta > 3$. Then for $N \geq 1$, in the regular case,*

$$\|\partial_\lambda^k K_j^\pm(\lambda)\|_{L^1 \to L^\infty} \leq C(1+\lambda)^{-2}, \quad \lambda \geq 0, \quad k = 0,1, \quad j = 1,2,3. \tag{7.2.40}$$

Proof We omit the signs $\pm$ again.

(i) Note that (7.2.40) for K_1 with $k = 0$ is exactly (7.2.35) with $k = 1$. Further, estimates (7.2.34), (7.2.30), and (7.2.36) imply

$$|\langle f, \partial_\lambda K_1(\lambda) g \rangle|$$
$$\leq \|V \partial_\lambda R_0^*(\lambda^2) f\|_{L^2_{\frac{3}{2}+}} \|\partial_\lambda \Pi(\lambda)\|_{L^2_{\frac{3}{2}+} \to L^2_{-\frac{3}{2}-}} \|V R_0(\lambda^2) g\|_{L^2_{\frac{3}{2}+}}$$
$$+ \|V R_0^*(\lambda^2) f\|_{L^2_{\frac{3}{2}+}} \|\partial_\lambda \Pi_N(\lambda)\|_{L^2_{\frac{3}{2}+} \to L^2_{-\frac{3}{2}-}} \|V \partial_\lambda R_0(\lambda^2) g\|_{L^2_{\frac{3}{2}+}}$$
$$+ \|V R_0^*(\lambda^2) f\|_{L^2_{\frac{5}{2}+}} \|\partial_\lambda^2 \Pi_N(\lambda)\|_{L^2_{\frac{5}{2}+} \to L^2_{-\frac{5}{2}-}} \|V R_0(\lambda^2) g\|_{L^2_{\frac{5}{2}+}}$$
$$\leq C(1+\lambda)^{-2} \|f\|_{L^1} \|g\|_{L^1}.$$

(ii) Now we estimate $\partial_\lambda^k K_2(\lambda)$. Note that

$$|\partial_\lambda^k G(\lambda, x, y)| \le |x|^k/(4\pi), \quad k = 0, 1, 2, \ldots \qquad (7.2.41)$$

Then, similarly to (7.2.36), we obtain for $0 \le \sigma \le \beta - k - 3/2$,

$$\|V\partial_\lambda^k G(\lambda)f\|_{L_\sigma^2} \le \|f\|_{L^1}, \quad k = 0, 1. \qquad (7.2.42)$$

Hence, (7.2.34), (7.2.30), and (7.2.36) imply

$$|\langle f, K_2(\lambda)g\rangle| \le \|VG(\lambda^2)f\|_{L^2_{\frac{1}{2}+}} \|\Pi(\lambda)\|_{L^2_{\frac{1}{2}+} \to L^2_{-\frac{1}{2}-}} \|VR_0(\lambda^2)g\|_{L^2_{\frac{1}{2}+}}$$

$$\le C(1+\lambda)^{-2}\|f\|_{L^1}\|g\|_{L^1}.$$

$$|\langle f, \partial_\lambda K_2(\lambda)g\rangle| \le \|V\partial_\lambda G(\lambda^2)f\|_{L^2_{\frac{3}{2}+}} \|\Pi(\lambda)\|_{L^2_{\frac{3}{2}+} \to L^2_{-\frac{3}{2}-}} \|VR_0(\lambda^2)g\|_{L^2_{\frac{3}{2}+}}$$

$$+ \|VG(\lambda^2)f\|_{L^2_{\frac{1}{2}+}} \|\Pi(\lambda)\|_{L^2_{\frac{1}{2}+} \to L^2_{-\frac{1}{2}-}} \|V\partial_\lambda R_0(\lambda^2)g\|_{L^2_{\frac{1}{2}+}}$$

$$+ \|VG(\lambda^2)f\|_{L^2_{\frac{3}{2}+}} \|\partial_\lambda \Pi(\lambda)\|_{L^2_{\frac{3}{2}+} \to L^2_{-\frac{3}{2}-}} \|VR_0(\lambda^2)g\|_{L^2_{\frac{3}{2}+}}$$

$$\le C(1+\lambda)^{-2}\|f\|_{L^1}\|g\|_{L^1}. \qquad (7.2.43)$$

The estimates for K_3 can be obtained similarly. $\qquad\square$

Corollary 7.2.10 *For $k = 0, 1$ the integral kernels $\partial_\lambda^k K_j(\lambda, x, y)$ belong to $L^\infty(\mathbb{R}^6)$, and*

$$\|\partial_\lambda^k K_j(\lambda, \cdot, \cdot)\|_{L^\infty(\mathbb{R}^6)} \le C(1+\lambda)^{-2}, \quad j = 1, 2, 3, \quad \lambda \ge 0.$$

Proof The distributional kernel $A(x, y)$ of any bounded linear operator $A \colon L^1(\mathbb{R}^3) \to L^\infty(\mathbb{R}^3)$ belongs to $L^\infty(\mathbb{R}^6)$, and

$$\|A(\cdot, \cdot)\|_{L^\infty(\mathbb{R}^6)} = \|A\|_{L^1(\mathbb{R}^3) \to L^\infty(\mathbb{R}^3)}.$$

This follows from the estimate $|\langle A, \phi\rangle| \le C\|\phi\|_{L^1(\mathbb{R}^6)}$ for $\phi \in L^1(\mathbb{R}^6)$ and from the duality $(L^1(\mathbb{R}^6))^* = L^\infty(\mathbb{R}^6)$. $\qquad\square$

Applying Corollary 7.2.10, we obtain

$$\sup_{x, y} |Q^\pm(t, x, y)| \le C|t|^{-\frac{1}{2}} \int_0^\infty (1 + |\lambda|)^{-2}d\lambda \le C_1|t|^{-\frac{1}{2}}, \quad |t| \ge 1$$

by the Van der Corput lemma. $\qquad\square$

8

Attractors and Quantum Mechanics

In this chapter we discuss possible relations of the foregoing theory of global attractors of nonlinear Hamiltonian equations to some mathematical problems of Quantum Theory.

This theory was inspired by fundamental postulates of quantum theory, primarily by Bohr's postulate on transitions between *quantum stationary states*. As a result, we have introduced a general summarizing conjecture (6). Here we discuss the possible relation of this conjecture to dynamical treatment of the Bohr postulates in the context of semiclassical nonlinear Maxwell–Schrödinger and Maxwell–Dirac equations.

8.1 Bohr's Postulates

In 1913 N. Bohr suggested the following two postulates which give the "Columbus solution" of the problem of stability and radiation of atoms and molecules [218]:

> **B1c.** *Electrons in atoms and molecules are permanently on some stationary orbits* $|E_n\rangle$ *with energies* E_n, *and sometimes make transitions between the orbits,*
>
> $$|E_n\rangle \mapsto |E_{n'}\rangle.$$
> (8.1.1)

> **B2.** *Such transition is followed by radiation of an electromagnetic wave of frequency* $\omega_{nn'} = \omega_{n'} - \omega_n$, *where* $\omega_k = E_k/\hbar$.
> (8.1.2)

Both of these postulates should have become theorems after the quantum theory of E. Schrödinger and W. Heisenberg emerged in 1925–1926. However, this did not happen till now, although both postulates are still actively used in quantum theory. This lack of theoretical clarity hinders progress in the theory (e.g., in superconductivity and in nuclear reactions), and in numerical simulation of many engineering processes (e.g., of laser radiation and quantum amplifiers) since a computer can solve dynamical equations but cannot take postulates into account.

The juxtaposition of the quantum postulates (8.1.1) and (8.1.2) with the Schrödinger theory raises the following questions.

I. Why are quantum stationary states (or *quantum stationary orbits*) in the Schrödinger theory identified with wave functions of the form (3.1.11)

$$\psi(x,t) = \psi_\omega(x)e^{-i\omega t} \ ? \tag{8.1.3}$$

II. Do Bohr's transitions (8.1.1) between these *quantum stationary states* allow a dynamical description?

Note that equation (8.1.3) implies that the amplitudes $\psi_\omega(x)$ are eigenfunctions of the Schrödinger operator.

The same questions arise in other dynamical models: quantum field theory, chromodynamics, and so on. However, the answer was not found till now.

The theory of attractors of Hamiltonian nonlinear PDEs, presented in this book, suggests that

(i) the form (8.1.3) of quantum stationary states is due to the $U(1)$-symmetry of the Schrödinger theory; that is, of the coupled Maxwell–Schrödinger equations (8.2.1),

(ii) the transitions can be interpreted as the global attraction (3.1.2) of all trajectories of a quantum system to an attractor formed by *stationary orbits* of type (8.1.3).

Moreover, the amplitudes of these stationary orbits are solutions of the *nonlinear eigenvalue problem* of type (3.1.12), which is *approximately linear* in a variety of cases due to the smallness of the interaction constant (the Sommerfeld constant).

We expect that other fundamental postulates of Quantum Theory also allow suitable interpretation in the framework of the theory of attractors for nonlinear Hamiltonian PDEs: *wave–particle duality* (L. de Broglie, 1924), and *probabilistic interpretation* (M. Born, 1927). More details can be found in [214].

8.2 On Dynamical Interpretation of Quantum Jumps

The simplest dynamical interpretation of the postulate **B1** is the global attraction to stationary orbits (3.1.2) for all finite-energy quantum trajectories $\psi(t)$. This means that stationary orbits form a global attractor of the corresponding quantum dynamics. However, this global attraction to stationary orbits contradicts the linear Schrödinger equation due to the superposition principle. Thus, Bohr's quantum jumps **B1** in the linear theory do not exist.

It is natural to suggest that the global attraction to stationary orbits (3.1.2) holds for a nonlinear modification of the linear Schrödinger theory. On the other hand, it turns out that even the original Schrödinger theory is nonlinear, because it involves interaction with the Maxwell field. The corresponding *semiclassical* nonlinear Maxwell–Schrödinger system was introduced essentially in the first of Schrödinger's articles [219] (see also Sections 4.2 and 12.4.2 of [213]):

$$\left\{ \begin{aligned} & i\hbar\dot\psi(x,t) = \frac{1}{2m}\left[-i\hbar\nabla - \frac{e}{c}(\mathbf{A}(x,t) + \mathbf{A}^{\text{ext}}(x,t)) \right]^2 \psi + e\big[A_0(x,t) \\ & \hspace{6cm} + A_0^{\text{ext}}(x,t)\big]\psi \\ & \Box A_\nu(x,t) = J_\nu(x,t), \qquad \nu = 0,1,2,3 \end{aligned} \right. ,$$

$$(8.2.1)$$

where $\Box$ is the d'Alembert wave operator $\frac{1}{c^2}\partial_t^2 - \Delta$. The Maxwell equations are written here in the four-dimensional form and in *unrationalized Gaussian units* (cgs) (or Heaviside–Lorentz units). The physical constants in these units are approximately equal to

$$e = -4.8 \times 10^{-10}\text{esu}, \ \ m = 9.1 \times 10^{-28}\text{g},$$
$$\hbar = 1.1 \times 10^{-27}\text{erg·s}, \ \ c = 3.0 \times 10^{10}\text{cm/s}$$

$$(8.2.2)$$

(see [212, p. 781] and [220, p. 221]). Further, $A = (A_0, \mathbf{A}) = (A_0, A_1, A_2, A_3)$ denotes four-dimensional potential of the Maxwell field in the Lorentz gauge $\dot A_0/c + \nabla \cdot \mathbf{A} = 0$, $A^{\text{ext}} = (A_0^{\text{ext}}, \mathbf{A}^{\text{ext}})$ is an external 4-potential, and J is the four-dimensional current density. To make these equations a closed system, we must also express the density of charge and currents via the wave function:

$$J_0(x,t) = e|\psi(x,t)|^2;$$
$$J_k(x,t) = \frac{e}{cm}[(-i\nabla_k + A_k(x,t) + A_k^{\text{ext}}(x,t))\psi(x,t)] \cdot \psi(x,t)$$

$$(8.2.3)$$

for $k = 1,2,3$, and "·" denotes the scalar product of two-dimensional real vectors corresponding to complex numbers. In particular, these expressions provide the continuity equation $\dot\rho + \operatorname{div} j = 0$ for any solution of the Schrödinger equation with arbitrary real potentials [213, Section 3.4].

System (8.2.1) is nonlinear in (ψ, A) although the Schrödinger equation is formally linear in ψ. It can be written as (5) in the case of *static external potentials*

$$\mathbf{A}^{\text{ext}}(x,t) \equiv \mathbf{A}^{\text{ext}}(x), \qquad A_0^{\text{ext}}(x,t) \equiv A_0^{\text{ext}}(x). \tag{8.2.4}$$

In this case the system (8.2.1) is G-invariant with the symmetry group $G = U(1)$. The symmetry means that for any solution $(\psi(x,t), A(x,t))$ of (8.2.1) and any $\theta \in \mathbb{R}$, the functions

$$T(e^{i\theta})(\psi(x,t),\ A(x,t)) := (\psi(x,t)e^{i\theta},\ A(x,t)) \tag{8.2.5}$$

are also solutions that can be easily verified. In particular, the 4-current (8.2.3) is invariant under this action. Now the "stationary G-orbits" (3.11.4) for the nonlinear hyperbolic system (8.2.1) are solutions of type

$$(\psi(x)e^{-i\omega t},\ A(x)). \tag{8.2.6}$$

The same remarks apply to the Maxwell–Dirac system introduced by Dirac in 1927:

$$\begin{cases} \displaystyle\sum_{\nu=0}^{3} \gamma^{\nu}[i\nabla_{\nu} - A_{\nu}(x,t) - A_{\nu}^{\text{ext}}(x,t)]\psi(x,t) = m\psi(x,t) \\[2mm] \Box A_{\nu}(x,t) = J_{\nu}(x,t) := \overline{\psi(x,t)}\gamma^{0}\gamma_{\nu}\psi(x,t), \quad \nu = 0, \ldots, 3 \end{cases} \Bigg| \; x \in \mathbb{R}^3, \tag{8.2.7}$$

where $\nabla_0 := \partial_t$.

We suggest that Bohr's quantum jumps **B1** for the systems (8.2.1) and (8.2.7) with a static external potential (8.2.4) can be interpreted as the *single-frequency asymptotics*

$$(\psi(x,t),\ A(x,t)) \sim (\psi_{\pm}(x)e^{-i\omega_{\pm}t},\ A_{\pm}(x)), \qquad t \to \pm\infty \tag{8.2.8}$$

for every finite-energy solution, where the asymptotics hold in local energy norms. These asymptotics correspond to our general conjecture (6) with the symmetry group $G = U(1)$ and its representation (8.2.5).

Stationary G-orbits (8.2.6) are solutions to the *nonlinear eigenvalue problem*

$$\begin{cases} \hbar\omega\psi(x) = \dfrac{1}{2m}\left[-i\hbar\nabla - \dfrac{e}{c}(\mathbf{A}(x) + \mathbf{A}^{\text{ext}}(x))\right]^2 \psi(x) + e\big[A_0(x) \\[3mm] \hspace{6cm} + A_0^{\text{ext}}(x)\big]\psi(x). \\[3mm] -\Delta A_{\nu}(x) = 4\pi J_{\nu}(x), \qquad \nu = 0,1,2,3 \end{cases} \tag{8.2.9}$$

The existence of these stationary G-orbits for the Maxwell–Schrödinger equations was established by G.M. Coclite and V. Georgiev [26] for the case of Coulomb external potentials

$$A_0^{\text{ext}} = -\frac{eZ}{|x|}, \qquad \mathbf{A}^{\text{ext}}(x) \equiv 0. \qquad (8.2.10)$$

For the Maxwell–Dirac system the existence of stationary G-orbits was established by M. Esteban, V. Georgiev, and E. Séré in the case of *zero external potentials* [27].

Remark 8.2.1 The *nonlinear eigenvalue problem* (8.2.9) reduces to the traditional linear eigenvalue problem for the Schrödinger operator if one neglects the "own Maxwell potentials" $A_0(x)$ and $\mathbf{A}(x)$ in the first equation. The solution of this linear eigenvalue problem with the normalization

$$\int |\psi(x)|^2 dx = 1 \qquad (8.2.11)$$

can be considered as the first approximation. Further one can apply perturbation procedure solving the Poisson equations in (8.2.9) with currents (8.2.3) defined with the first approximation, adding their solutions to the external potentials, and so on. The convergence of this procedure is not proved, though it gives satisfactory results in a variety of cases.

Furthermore, in the case of *zero external potentials* the Maxwell–Schrödinger system is translation-invariant, while the Maxwell–Dirac system is relativistic. Respectively, for their solutions one should expect the soliton asymptotics of type (13) in global energy norms as $t \to \pm\infty$:

$$\left\{ \begin{array}{l} \psi(x,t) \sim \sum_k \psi_{\pm}^k(x - v_{\pm}^k t) e^{i\Phi_{\pm}^k(x,t)} + \varphi_{\pm}(x,t), \\ A(x,t) \sim \sum_k A_{\pm}^k(x - v_{\pm}^k t) + A_{\pm}(x,t). \end{array} \right. \qquad (8.2.12)$$

Here $\Phi_{\pm}^k(x,t)$ are suitable phase functions, and each soliton ($\psi_{\pm}^k(x - v_{\pm}^k t)e^{i\Phi_{\pm}^k(x,t)}$, $A_{\pm}^k(x - v_{\pm}^k t)$) is a solution of the corresponding "nonlinear eigenvalue problem" of type (3.1.12), while $\varphi_{\pm}(x,t)$ and $A_{\pm}(x,t)$ represent some dispersive waves which are solutions to the free Schrödinger and Maxwell equations respectively. The asymptotics (8.2.12) provisionally describe the wave–particle duality suggested by L. de Broglie in his PhD thesis of 1923 [213, 214].

The asymptotics (8.2.8) and (8.2.12) are not proved yet for the Maxwell–Schrödinger and Maxwell–Dirac systems (8.2.1) and (8.2.7). One might expect that these asymptotics should follow by suitable modification of the arguments from Chapter 3, which give a rigorous justification of similar arguments for $U(1)$-invariant equations (3.1.1) and (3.2.1)–(3.2.3). However, a rigorous justification for the systems (8.2.1) and (8.2.7) is still an open problem.

Remark 8.2.2 The Schrödinger definition of quantum stationary states (8.1.3) has been formalized by J. von Neumann [215] saying that quantum states are the rays in the Hilbert space, which is just a reformulation of (8.1.3). Our main conjecture is that these definitions are indeed a **theorem**: namely, we suggest that the asymptotics (8.2.8) are inherent properties of the nonlinear Maxwell–Schrödinger equations (8.2.1). This conjecture is confirmed by perturbative arguments below.

8.3 Bohr's Postulates via Perturbation Theory

The remarkable success of the Schrödinger theory was the explanation of Bohr's postulates in the case of *static external potentials* by *perturbation theory* applied to the *coupled Maxwell–Schrödinger equations* (8.2.1). Namely, as a first approximation, the time-dependent fields $\mathbf{A}(x,t)$ and $A^0(x,t)$ in the Schrödinger equation of the system (8.2.1) can be neglected:

$$i\hbar\dot{\psi}(x,t) = H\psi(x,t) := \frac{1}{2m}\left[-i\hbar\nabla - \frac{e}{c}\mathbf{A}_{\text{ext}}(x)\right]^2\psi(x,t) + eA^0_{\text{ext}}(x)\psi(x,t).$$

$$(8.3.1)$$

For "sufficiently good" external potentials and initial conditions any finite-energy solution can be expanded in eigenfunctions

$$\psi(x,t) = \sum_n C_n\psi_n(x)e^{-i\omega_n t} + \psi_c(x,t), \qquad \psi_c(x,t) = \int C(\omega)e^{-i\omega t}d\omega,$$

$$(8.3.2)$$

where integration is performed over the continuous spectrum of the Schrödinger operator H, and for any $R > 0$,

$$\int_{|x|<R} |\psi_c(x,t)|^2 dx \to 0, \quad t \to \pm\infty, \qquad (8.3.3)$$

see, for example, [188, Theorem 21.1]. The substitution of this expansion into the expression for current density (8.2.3) gives the series

$$J(x,t) = \sum_{nn'} J_{nn'}(x)e^{-i\omega_{nn'}t} + c.c. + J_c(x,t), \qquad (8.3.4)$$

where $J_c(x,t)$ has a continuous frequency spectrum. Therefore, the currents on the right-hand side of the Maxwell equation from (8.2.1) contain, besides the continuous spectrum, only discrete frequencies $\omega_{nn'}$. Hence, the discrete spectrum of the corresponding Maxwell field also contains only these frequencies $\omega_{nn'}$. This proves Bohr's rule **B2** *in the first order of perturbation theory* only, since this calculation ignores the inverse effect of radiation onto the atom.

Moreover, these arguments also suggest that we treat the jumps (8.1.1) as the *single-frequency asymptotics* (8.2.8) for solutions to the Schrödinger equation *coupled to the Maxwell equations*.

Namely, the currents (8.3.4) on the right of the Maxwell equation from (8.2.1) produce the radiation when nonzero frequencies $\omega_{nn'}$ are present. This is due to the fact that $\mathbb{R} \setminus 0$ is a subset of absolutely continuous spectrum of the Maxwell equations.

However, this radiation cannot last forever, since it irrevocably carries the energy to infinity while the total energy is finite. Hence in the long-time limit only $\omega_{nn'} = 0$ should survive, which means exactly the single-frequency asymptotics (8.2.8) by (8.3.3).

Remark 8.3.1 Of course, these perturbation arguments cannot provide a rigorous justification of the long-time asymptotics (4.5.21) for the coupled Maxwell–Schrödinger equations. In [60]–[71], we have justified similar single-frequency asymptotics for a list of model $U(1)$-invariant nonlinear PDEs, see Chapter 3. Nevertheless, for the coupled Maxwell–Schrödinger equation such justification is still an open problem.

8.4 Conclusion

The discussion above suggests that N. Bohr's postulates cannot be interpreted in the framework of the linear Schrödinger equation alone, but admit a hypothetical explanation in the framework of the coupled Maxwell–Schrödinger equations. In [214] we also suggested a mathematical treatment of other fundamental postulates of Quantum Theory relying on the coupled Maxwell–Schrödinger equations: of L. de Broglie's wave–particle duality and of M. Born's probabilistic interpretation.

It seems, the absence of a suitable treatment of these postulates in the framework of linear theory was the cause of heated discussions between A. Einstein and N. Bohr together with other physicists [205]. Note that W. Heisenberg began developing a nonlinear theory of elementary particles [208, 209].

According to many expert physicists, a mathematical analysis of problems of quantum mechanics is useless because its area of applicability is limited, as are its capabilities. However, the purpose of our discussions is not in improving the physical theory. Our goal is to prepare a mathematical ground for approach to some open questions of Quantum Theory which are not accessible with perturbation technique, for instance, to the questions of nuclei classification and of nuclear reactions. We suppose that the nuclei are "quantum stationary states" of suitable nonlinear equations, i.e., points of the corresponding global attractor.

Note, the second-quantized Maxwell–Schrödinger system is the main subject of quantum electrodynamics [217]. Our specific attention to the *semiclassical* Maxwell–Schrödinger and Maxwell–Dirac systems is due to the fact that for these systems there is an extensive empirical material: on atomic spectra, electron diffraction, on crystals and their thermal and electric conductivity, etc. Therefore, one can try to find a possible mathematical description of these phenomena in the framework of these systems. So these semiclassical systems serve as a testing ground for a development of the mathematical theory.

Similar questions also exist on a higher level in the quantum field theory [217]. However, they obviously cannot be clarified until these questions are understood in a simpler context of semiclassical theory.

Bibliography

Differential Equations and Functional Analysis

[1] V. I. Arnold, Mathematical Methods of Classical Mechanics, Springer, New York, 1989.

[2] T. Cazenave, Semilinear Schrödinger Equations, AMS, New York, 2003.

[3] T. Cazenave, A. Haraux, An Introduction to Semilinear Evolution Equations, Clarendon Press, Oxford, 1998.

[4] G. I. Gaudry, Quasimeasures and operators commuting with convolution, *Pacific J. Math.* **18** (1966), 461–476.

[5] E. Hopf, Über die Anfangswertaufgabe für die hydrodynamischen Grundgleichungen, *Math. Nachr.* **4** (1951), 213–231.

[6] L. Hörmander, The Analysis of Linear Partial Differential Operators, *vol. I, 2nd ed.*, Springer, Berlin, 1990.

[7] K. Jörgens, Das Anfangswertproblem im Grossen für eine Klasse nichtlinearer Wellengleichungen, *Math. Z.* **77** (1961), 295–308.

[8] A. I. Komech, Attractors of nonlinear Hamiltonian PDEs, *Discrete Continuous Dyn. Syst. A* **36** (2016), 6201–6256. arXiv:1409.2009

[9] A. I. Komech, A. A. Komech, Principles of Partial Differential Equations, Springer, Berlin, 2009.

[10] A. I. Komech, A. E. Merzon, Stationary Diffraction by Wedges: Method of Automorphic Functions on Complex Characteristics, Lecture Notes in Mathematics 2249, Springer, Berlin, 2019.

[11] A. I. Komech, E. A. Kopylova, Attractors of Hamiltonian nonlinear partial differential equations, *Russ. Math. Surv.* **75** (2020), 1–87.

[12] J.-L. Lions, Quelques Méthodes de Résolution des Problèmes aux Limites non Linéaires, Dunod, Gauthier–Villars, Paris, 1969.

[13] M. Reed, B. Simon, Methods of Modern Mathematical Physics, Academic Press, New York, I (1980), II (1975), III (1979), IV (1978).

[14] F. Riesz, B. Sz.-Nagy, Functional Analysis, Dover, New York, 1990.

[15] W. Rudin, Functional Analysis, McGraw-Hill, New York, 1977.

Global Attractors of Dissipative PDEs

[16] A. V. Babin, M. I. Vishik, Attractors of Evolution Equations, Studies in Mathematics and Its Applications 25, North-Holland, Amsterdam, 1992.

[17] V. V. Chepyzhov, M. I. Vishik, Attractors for Equations of Mathematical Physics, American Mathematical Society Colloquium Publications 49, American Mathematical Society, Providence, RI, 2002.

[18] C. Foias, O. Manley, R. Rosa, R. Temam, Navier-Stokes Equations and Turbulence, Encyclopedia of Mathematics and Its Applications 83, Cambridge University Press, Cambridge, 2001.

[19] J. K. Hale, Asymptotic Behavior of Dissipative Systems, Mathematical Surveys and Monographs 25, American Mathematical Society, Providence, RI, 1988.

[20] A. Haraux, Systémes Dynamiques Dissipatifs et Applications, R.M.A. 17, Collection dirigé par Ph. Ciarlet et J.-L. Lions, Masson, Paris, 1990.

[21] D. Henry, Geometric Theory of Semilinear Parabolic Equations, Lecture Notes in Mathematics 480, Springer, Berlin, 1981.

[22] L. Landau, On the problem of turbulence, *C. R. (Doklady) Acad. Sci. URSS (N.S.)* **44** (1944), 311–314.

[23] R. Temam, Infinite-Dimensional Dynamical Systems in Mechanics and Physics, Springer, New York, 1997.

The Existence of Stationary Orbits and Solitons

[24] H. Berestycki, P.-L. Lions, Nonlinear scalar field equations. I. Existence of a ground state, *Arch. Rational Mech. Anal.* **82** (1983), 313–345.

[25] H. Berestycki, P.-L. Lions, Nonlinear scalar field equations. II. Existence of infinitely many solutions, *Arch. Rational Mech. Anal.* **82** (1983), 347–375.

[26] G. M. Coclite, V. Georgiev, Solitary waves for Maxwell–Schrödinger equations, *Electron. J. Differential Equations* **94** (2004), 31 pp.

[27] M. J. Esteban, V. Georgiev, E. Séré, Stationary solutions of the Maxwell-Dirac and the Klein–Gordon–Dirac equations, *Calc. Var. Partial Differential Equations* **4** (1996), 265–281.

[28] L. Lusternik, L. Schnirelmann, Méthodes Topologiques Dans les Problèmes Variationels, Hermann, Paris, 1934.

[29] L. Lusternik, L. Schnirelmann, Topological methods in variational problems and their applications to differential geometry of surfaces, *Uspekhi Mat. Nauk* **2** (1947), 166–217.

[30] W. A. Strauss, Existence of solitary waves in higher dimensions, *Comm. Math. Phys.* **55** (1977), 149–162.

Local Energy Decay

[31] P. D. Lax, C. S. Morawetz, R. S. Phillips, Exponential decay of solutions of the wave equation in the exterior of a star-shaped obstacle, *Comm. Pure Appl. Math.*, **16** (1963), 477–486.

[32] C. S. Morawetz, Time decay for the nonlinear Klein–Gordon equations, *Proc. R. Soc. London Ser. A* **306** (1968), 291–296.

[33] C. S. Morawetz, W. A. Strauss, Decay and scattering of solutions of a nonlinear relativistic wave equation, *Comm. Pure Appl. Math.* **25** (1972), 1–31.

[34] I. Segal, Quantization and dispersion for nonlinear relativistic equations, pp. 79–108 in: Mathematical Theory of Elementary Particles (Proc. Conf., Dedham, Mass., 1965), MIT Press, Cambridge, MA, 1966.

[35] I. Segal, Dispersion for non-linear relativistic equations. II, *Ann. Sci. École Norm. Sup. (4)* **1** (1968), 459–497.

[36] W. A. Strauss, Decay and asymptotics for $\Box u = F(u)$, *J. Funct. Anal.* **2** (1968), 409–457.

[37] B. R. Vainberg, Asymptotic Methods in Equations of Mathematical Physics, Gordon and Breach, New York, 1989.

Global Attraction to Stationary States for Hamiltonian PDEs

[38] M. Freidlin, A. I. Komech, On metastable regimes in stochastic Lamb system, *J. Math. Phys.* **47** (2006), 043301.

[39] J. B. Keller, L. L. Bonilla, Irreversibility and nonrecurrence, *J. Stat. Physics* **42** (1986), 1115–1125.

[40] A. I. Komech, On the stabilization of interaction of a string with a nonlinear oscillator, *Moscow Univ. Math. Bull.* **46**, no. 6 (1991), 34–39.

[41] A. I. Komech, On stabilization of string-nonlinear oscillator interaction, *J. Math. Anal. Appl.* **196** (1995), 384–409.

[42] A. I. Komech, On the stabilization of string-oscillator interaction, *Russ. J. Math. Phys.* **3** (1995), 227–247.

[43] A. I. Komech, On transitions to stationary states in one-dimensional nonlinear wave equations, *Arch. Ration. Mech. Anal.* **149** (1999), 213–228.

[44] A. I. Komech, H. Spohn, M. Kunze, Long-time asymptotics for a classical particle interacting with a scalar wave field, *Comm. Partial Differential Equations* **22** (1997), 307–335.

[45] A. I. Komech, H. Spohn, Long-time asymptotics for the coupled Maxwell–Lorentz equations, *Comm. Partial Differential Equations* **25** (2000), 559–584.

[46] A. Komech, Attractors of non-linear Hamiltonian one-dimensional wave equations, *Russ. Math. Surv.* **55**, no. 1 (2000), 43–92.

[47] A. I. Komech, A. Merzon, Scattering in the nonlinear Lamb system, *Phys. Lett. A* **373** (2009), 1005–1010.

[48] A. I. Komech, A. Merzon, On asymptotic completeness for scattering in the nonlinear Lamb system, *J. Math. Phys.* **50** (2009), 023514.

[49] A. I. Komech, A. Merzon, On asymptotic completeness of scattering in the nonlinear Lamb system, II, *J. Math. Phys.* **54** (2013), 012702.

[50] E. Kopylova, On global attraction to stationary states for wave equation with concentrated nonlinearity, *J. Dyn. Diff. Equations* **30**, no. 1 (2018), 107–116.

[51] H. Lamb, On a peculiarity of the wave-system due to the free vibrations of a nucleus in an extended medium, *Proc. London Math. Soc.* **32** (1900), 208–211.

[52] H. Spohn, Dynamics of Charged Particles and Their Radiation Field, Cambridge University Press, Cambridge, 2004.

Global Attraction to Solitons for Hamiltonian PDEs

[53] W. Eckhaus, A. van Harten, The Inverse Scattering Transformation and the Theory of Solitons, North-Holland, Amsterdam, 1981.

[54] A. I. Komech, H. Spohn, Soliton asymptotics for a classical particle interacting with a scalar wave field, *Nonlinear Anal.* **33** (1998), 13–24.

[55] V. Imaykin, A. I. Komech, N. Mauser, Soliton-type asymptotics for the coupled Maxwell-Lorentz equations, *Ann. Henri Poincaré* **P** (2004), 1117–1135.

[56] V. Imaykin, A. I. Komech, H. Spohn, Soliton-type asymptotics and scattering for a charge coupled to the Maxwell field, *Russ. J. Math. Phys.* **9** (2002), 428–436.

[57] V. Imaykin, A. I. Komech, H. Spohn, Scattering theory for a particle coupled to a scalar field, *Discrete Continuous Dyn. Syst.* **10** (2004), 387–396.

[58] A. I. Komech, N. J. Mauser, A. P. Vinnichenko, Attraction to solitons in relativistic nonlinear wave equations, *Russ. J. Math. Phys.* **11** (2004), 289–307.

[59] G. L. Lamb Jr., Elements of Soliton Theory, John Wiley, New York, 1980.

Global Attraction to Stationary Orbits for Hamiltonian PDEs

[60] A. Comech, Weak attractor of the Klein–Gordon field in discrete space-time interacting with a nonlinear oscillator, *Discrete Continuous Dyn. Syst.* **33** (2013), 2711–2755.

[61] A. Comech, On global attraction to solitary waves: Klein–Gordon equation with mean field interaction at several points, *J. Differential Equations* **252** (2012), 5390–5413.

[62] A. I. Komech, On attractor of a singular nonlinear U(1)-invariant Klein–Gordon equation, pp. 599–611 in: Progress in Analysis, World Scientific, River Edge, NJ, 2003.

[63] A. I. Komech, A. A. Komech, On the global attraction to solitary waves for the Klein–Gordon equation coupled to a nonlinear oscillator, *C. R. Math. Acad. Sci. Paris* **343** (2006), 111–114.

[64] A. I. Komech, A. A. Komech, Global attractor for a nonlinear oscillator coupled to the Klein–Gordon field, *Arch. Ration. Mech. Anal.* **185** (2007), 105–142.

[65] A. I. Komech, A. A. Komech, On global attraction to solitary waves for the Klein–Gordon field coupled to several nonlinear oscillators, *J. Math. Pure Appl.* **93** (2010), 91–111.

[66] A. I. Komech, A. A. Komech, Global attraction to solitary waves for Klein–Gordon equation with mean field interaction, *Ann. Inst. H. Poincaré Anal. Non Linéaire* **26** (2009), 855–868.

[67] A. I. Komech, A. A. Komech, Global attraction to solitary waves for a nonlinear Dirac equation with mean field interaction, *SIAM J. Math. Anal.* **42** (2010), 2944–2964.

[68] A. A. Komech, A. I. Komech, A variant of the Titchmarsh convolution theorem for distributions on the circle, *Funktsional. Anal. Prilozhen.* **47** (2013), 26–32.

[69] E. Kopylova, Global attraction to solitary waves for Klein–Gordon equation with concentrated nonlinearity, *Nonlinearity* **30**, no. 11 (2017), 4191–4207.

[70] E. Kopylova, A. I. Komech, On global attractor of 3D Klein–Gordon equation with several concentrated nonlinearities, *Dyn. PDE* **16**, no. 2 (2019), 105–124.

[71] E. Kopylova, A. I. Komech, Global attractor for 1D Dirac field coupled to nonlinear oscillator, *Comm. Math. Phys.* **375**, no. 1 (2020), 573–603.

[72] O. A. Ladyženskaya, On the limiting amplitude principle, *Uspekhi Mat. Nauk* **12** (1957), 161–164. [Russian]

[73] B. Y. Levin, Lectures on Entire Functions, American Mathematical Society, Providence, RI, 1996.

[74] L. Lewin, Advanced Theory of Waveguides, Iliffe, London, 1951.

[75] C. S. Morawetz, The limiting amplitude principle, *Comm. Pure Appl. Math.* **15** (1962), 349–361.

[76] I. M. Sigal, Nonlinear wave and Schrödinger equations. I. Instability of periodic and quasiperiodic solutions, *Comm. Math. Phys.* **153** (1993), 297–320.

[77] W. A. Strauss, Nonlinear scattering theory at low energy, *J. Funct. Anal.* **41** (1981), 110–133.

[78] W. A. Strauss, Nonlinear scattering theory at low energy: Sequel, *J. Funct. Anal.* **43** (1981), 281–293.

[79] E. C. Titchmarsh, The zeros of certain integral functions, *Proc. London Math. Soc.* **S2–25**, no. 1 (1926), 283–302.

Adiabatic Effective Dynamics of Solitons

[80] V. Bach, T. Chen, J. Faupin, J. Fröhlich, I. M. Sigal, Effective dynamics of an electron coupled to an external potential in non-relativistic QED, *Ann. Henri Poincaré* **14** (2013), 1573–1597.

[81] S. Demoulini, D. Stuart, Adiabatic limit and the slow motion of vortices in a Chern–Simons–Schrödinger system, *Comm. Math. Phys.* **290** (2009), 597–632.

[82] J. Fröhlich, S. Gustafson, B. L. G. Jonsson, I. M. Sigal, Solitary wave dynamics in an external potential, *Comm. Math. Phys.* **250** (2004), 613–642.

[83] J. Fröhlich, T.-P. Tsai, H.-T. Yau, On the point-particle (Newtonian) limit of the nonlinear Hartree equation, *Comm. Math. Phys.* **225** (2002), 223–274.

[84] A. Komech, M. Kunze, H. Spohn, Effective dynamics for a mechanical particle coupled to a wave field, *Comm. Math. Phys.* **203** (1999), 1–19.

[85] M. Kunze, H. Spohn, Adiabatic limit for the Maxwell–Lorentz equations, *Ann. Henri Poincaré* **1** (2000), 625–653.

[86] E. Long, D. Stuart, Effective dynamics for solitons in the nonlinear Klein–Gordon–Maxwell system and the Lorentz force law, *Rev. Math. Phys.* **21** (2009), 459–510.

[87] D. Stuart, Existence and Newtonian limit of nonlinear bound states in the Einstein–Dirac system, *J. Math. Phys.* **51** (2010), 032501, 13.

Global Attraction to Stationary $SO(3)$-Orbits

[88] V. Imaykin, A. I. Komech, H. Spohn, Rotating charge coupled to the Maxwell field: Scattering theory and adiabatic limit, *Monatsh. Math.* **142** (2004), 143–156.

Concentrated Nonlinearities

[89] R. Adami, D. Noja, C. Ortoleva, Orbital and asymptotic stability for standing waves of a nonlinear Schrödinger equation with concentrated nonlinearity in dimension three, *J. Math. Phys.* **54**, no. 1 (2013), 013501.

[90] S. Albeverio, R. Figari, Quantum fields and point interactions, *Rend. Mat. Appl. (7)* **39** (2018), 161–180.

[91] F. A. Berezin, L. D. Faddeev, Remark on the Schrödinger equation with singular potential, *Soviet Math. Dokl.* **2** (1961), 372–375.

[92] F. H. Cornish, Classical radiation theory and point charges, *Proc. Phys. Soc.* **86**, no 3 (1965), 427–442.

[93] P. A. M. Dirac, Classical theory of radiating electrons, *Proc. R. Soc. A* **167** (1938), 148–169.

[94] H.-P. Gittel, J. Kijowski, E. Zeidler, The relativistic dynamics of the combined particle-field system in renormalized classical electrodynamics, *Comm. Math. Phys.* **198** (1998), 711–736.

[95] D. Noja, A. Posilicano, Wave equations with concentrated nonlinearities. *J. Phys. A* **38** (2005), 5011–5022.

[96] D. R. Yafaev, On a zero-range interaction of a quantum particle with the vacuum, *J. Phys. A* **25** (1992), 963–978.

[97] D. R. Yafaev, A point interaction for the discrete Schrödinger operator and generalized Chebyshev polynomial, *J. Math. Phys.* **58**, no. 6 (2017), 063511.

[98] Ya. B. Zel'dovich, Scattering by a singular potential in perturbation theory and in the momentum representation, *JETP* **11** (1960), 594–597.

Orbital Stability of Solitons

[99] D. Bambusi, L. Galgani, Some rigorous results on the Pauli–Fierz model of classical electrodynamics, *Ann. H. Poincaré Phys. Theor.* **58** (1993), 155–171.

[100] M. Grillakis, J. Shatah, W. Strauss, Stability theory of solitary waves in the presence of symmetry. I, *J. Funct. Anal.* **74** (1987), 160–197.

[101] M. Grillakis, J. Shatah, W. Strauss, Stability theory of solitary waves in the presence of symmetry. II, *J. Funct. Anal.* **94** (1990), 308–348.

[102] J. E. Marsden, T. S. Ratiu, Introduction to Mechanics and Symmetry, Springer, Berlin, 1994.

[103] Y. G. Oh, A stability criterion for Hamiltonian systems with symmetry, *J. Geom. Phys.* **4** (1987), 163–182.

[104] H. Poincaré, Sur l'équilibre d'une masse fluide animée d'un mouvement de rotation, *Acta. Math.* **7** (1885), 259–380.

Asymptotic Stability of Solitary Manifolds

[105] L. Andersson, P. Blue, Uniform energy bound and asymptotics for the Maxwell field on a slowly rotating Kerr black hole exterior, *J. Hyperbolic Diff. Equations* **12** (2015), 689–743.

[106] D. Bambusi, S. Cuccagna, On dispersion of small energy solutions to the nonlinear Klein–Gordon equation with a potential, *Am. J. Math.* **133** (2011), 1421–1468.

[107] A. Bensoussan, C. Iliine, A. I. Komech, Breathers for a relativistic nonlinear wave equation, *Arch. Ration. Mech. Anal.* **165** (2002), 317–345.

[108] N. Boussaid, Stable directions for small nonlinear Dirac standing waves, *Comm. Math. Phys.* **268** (2006), 757–817.

[109] N. Boussaid, S. Cuccagna, On stability of standing waves of nonlinear Dirac equations, *Comm. Partial Diff. Equations* **37** (2012), 1001–1056.

[110] V. S. Buslaev, G. S. Perelman, Scattering for the nonlinear Schrödinger equation: States that are close to a soliton, *Algebra Anal.* **4** (1992), 63–102.

[111] V. S. Buslaev, G. S. Perelman, On the stability of solitary waves for nonlinear Schrödinger equations, pp. 75–98 in: Nonlinear Evolution Equations, Amer. Math. Soc. Transl. Ser. 2 164, AMS, Providence, RI, 1995.

[112] V. Buslaev, C. Sulem, On asymptotic stability of solitary waves for nonlinear Schrödinger equations, *Ann. Inst. H. Poincaré Anal. Non Linéaire* **20** (2003), 419–475.

[113] V. Buslaev, A. I. Komech, E. Kopylova, D. Stuart, On asymptotic stability of solitary waves in Schrödinger equation coupled to nonlinear oscillator, *Comm. Partial Diff. Equations* **33** (2008), 669–705.

[114] S. Cuccagna, Stabilization of solutions to nonlinear Schrödinger equations, *Comm. Pure Appl. Math.* **54** (2001), 1110–1145.

[115] S. Cuccagna, The Hamiltonian structure of the nonlinear Schrödinger equation and the asymptotic stability of its ground states, *Comm. Math. Phys.* **305** (2011), 279–331.

[116] S. Cuccagna, T. Mizumachi, On asymptotic stability in energy space of ground states for nonlinear Schrödinger equations, *Comm. Math. Phys.* **284** (2008), 51–77.

[117] M. Dafermos, I. Rodnianski, A proof of the uniform boundedness of solutions to the wave equation on slowly rotating Kerr backgrounds, *Invent. Math.* **185** (2011), 467–559.

[118] R. Donninger, W. Schlag, A. Soffer, On pointwise decay of linear waves on a Schwarzschild black hole background, *Comm. Math. Phys.* **309** (2012), 51–86.

[119] T. Duyckaerts, C. Kenig, F. Merle, Profiles of bounded radial solutions of the focusing, energy-critical wave equation, *Geom. Funct. Anal.* **22** (2012), 639–698.

[120] T. Duyckaerts, C. Kenig, F. Merle, Scattering for radial, bounded solutions of focusing supercritical wave equations, *Int. Math. Res. Not. IMRN* **2014**, no. 1 (2014), 224–258.

[121] T. Duyckaerts, C. Kenig, F. Merle, Concentration-compactness and universal profiles for the non-radial energy critical wave equation, *Nonlinear Anal.* **138** (2016), 44–82.

[122] J. Fröhlich, Z. Gang, Emission of Cherenkov radiation as a mechanism for Hamiltonian friction, *Adv. Math.* **264** (2014), 183–235.

[123] T. Harada, H. Maeda, Stability criterion for self-similar solutions with a scalar field and those with a stiff fluid in general relativity, *Classical Quantum Gravity* **21** (2004), 371–389.

[124] V. Imaikin, Soliton asymptotics for systems of field-particle type, *Russ. Math. Surv.* **68** (2013), 227–281.

[125] V. Imaykin, A. I. Komech, H. Spohn, Scattering asymptotics for a charged particle coupled to the Maxwell field, *J. Math. Phys.* **52** (2011), 042701.

[126] V. Imaykin, A. I. Komech, B. Vainberg, On scattering of solitons for the Klein–Gordon equation coupled to a particle, *Comm. Math. Phys.* **268** (2006), 321–367.

[127] V. Imaykin, A. I. Komech, B. Vainberg, Scattering of solitons for coupled wave-particle equations, *J. Math. Anal. Appl.* **389** (2012), 713–740.

[128] C. Kenig, A. Lawrie, B. Liu, W. Schlag, Stable soliton resolution for exterior wave maps in all equivariance classes, *Adv. Math.* **285** (2015), 235–300.

[129] C. E. Kenig, A. Lawrie, W. Schlag, Relaxation of wave maps exterior to a ball to harmonic maps for all data, *Geom. Funct. Anal.* **24** (2014), 610–647.

[130] C. E. Kenig, F. Merle, Global well-posedness, scattering and blow-up for the energy-critical, focusing, non-linear Schrödinger equation in the radial case, *Invent. Math.* **166** (2006), 645–675.

[131] C. E. Kenig, F. Merle, Global well-posedness, scattering and blow-up for the energy-critical focusing non-linear wave equation, *Acta Math.* **201** (2008), 147–212.

[132] C. E. Kenig, F. Merle, Nondispersive radial solutions to energy supercritical non-linear wave equations, with applications, *Am. J. Math.* **133** (2011), 1029–1065.

[133] A. I. Komech, A. A. Komech, Global well-posedness for the Schrödinger equation coupled to a nonlinear oscillator, *Russ. J. Math. Phys.* **14** (2007), 164–173.

[134] A. I. Komech, E. Kopylova, Scattering of solitons for the Schrödinger equation coupled to a particle, *Russ. J. Math. Phys.* **13** (2006), 158–187.

[135] E. Kopylova, A. I. Komech, On asymptotic stability of moving kink for relativistic Ginzburg–Landau equation, *Comm. Math. Phys.* **302** (2011), 225–252.

[136] E. Kopylova, A. I. Komech, On asymptotic stability of kink for relativistic Ginzburg–Landau equations, *Arch. Ration. Mech. Anal.* **202** (2011), 213–245.

[137] A. I. Komech, E. Kopylova, On eigenfunction expansion of solutions to the Hamiltonian equations, *J. Stat. Phys.* **154** (2014), 503–521.

[138] A. I. Komech, E. Kopylova, On the eigenfunction expansion for the Hamiltonian operators, *J. Spectr. Theory* **5** (2015), 331–361.

[139] A. I. Komech, E. Kopylova, S. Kopylov, On nonlinear wave equations with parabolic potentials, *J. Spectr. Theory* **3** (2013), 485–503.

[140] A. I. Komech, E. Kopylova, H. Spohn, Scattering of solitons for Dirac equation coupled to a particle, *J. Math. Anal. Appl.* **383** (2011), 265–290.

[141] A. I. Komech, E. Kopylova, D. Stuart, On asymptotic stability of solitons in a nonlinear Schrödinger equation, *Comm. Pure Appl. Anal.* **11** (2012), 1063–1079.

[142] E. Kopylova, On asymptotic stability of solitary waves in discrete Schrödinger equation coupled to nonlinear oscillator, *Nonlinear Anal. Ser. A Theory Methods Appl.* **71** (2009), 3031–3046.

[143] E. Kopylova, On asymptotic stability of solitary waves in discrete Klein–Gordon equation coupled to nonlinear oscillator, *Applicable Anal.* **89** (2010), 1467–1493.

[144] E. Kopylova, Asymptotic stability of solitons for non-linear hyperbolic equations, *Russ. Math. Surv.* **68** (2013), 283–334.

[145] M. G. Krein, H. K. Langer, The spectral function of a self-adjoint operator in a space with indefinite metric, *Sov. Math. Dokl.* **4** (1963), 1236–1239.

[146] J. Krieger, K. Nakanishi, W. Schlag, Center-stable manifold of the ground state in the energy space for the critical wave equation, *Math. Ann.* **361** (2015), 1–50.

[147] J. Krieger, W. Schlag, Concentration Compactness for Critical Wave Maps, EMS Monographs in Mathematics, European Mathematical Society, Zürich, 2012.

[148] H. Langer, Spectral functions of definitizable operators in Krein spaces, pp. 1–46 in: D. Butkovic, H. Kraljevic, S. Kurepa (eds.), Functional Analysis: Proceedings of a Conference Held in Dubrovnik, November 2–14, 1981, Lecture Notes in Mathematics 948, Springer, Berlin, 1982.

[149] Y. Martel, Asymptotic N-soliton-like solutions of the subcritical and critical generalized Korteweg–de Vries equations, *Am. J. Math.* **127** (2005), 1103–1140.

[150] Y. Martel, F. Merle, Asymptotic stability of solitons of the subcritical gKdV equations revisited, *Nonlinearity* **18** (2005), 55–80.

[151] Y. Martel, F. Merle, T. P. Tsai, Stability and asymptotic stability in the energy space of the sum of N solitons for subcritical gKdV equations, *Comm. Math. Phys.* **231** (2002), 347–373.

[152] M. Merkli, I. M. Sigal, A time-dependent theory of quantum resonances, *Comm. Math. Phys.* **201** (1999), 549–576.

[153] J. R. Miller, M. I. Weinstein, Asymptotic stability of solitary waves for the regularized long-wave equation, *Comm. Pure Appl. Math.* **49** (1996), 399–441.

[154] K. Nakanishi, W. Schlag, Invariant Manifolds and Dispersive Hamiltonian Evolution Equations, Zurich Lectures in Advanced Mathematics, European Mathematical Society, Zürich, 2011.

[155] C. A. Pillet, C. E. Wayne, Invariant manifolds for a class of dispersive, Hamiltonian, partial differential equations, *J. Diff. Equations* **141** (1997), 310–326.

[156] R. L. Pego, M. I. Weinstein, Asymptotic stability of solitary waves, *Comm. Math. Phys.* **164** (1994), 305–349.

[157] G. Perelman, Asymptotic stability of multi-soliton solutions for nonlinear Schrödinger equations, *Comm. Partial Diff. Equations* **29** (2004), 1051–1095.

[158] M. Reed, B. Simon, Methods of Modern Mathematical Physics. IV. Analysis of Operators, Academic Press, New York, 1978.

[159] I. Rodnianski, W. Schlag, A. Soffer, Asymptotic stability of N-soliton states of NLS, ArXiv: math/ 0309114.

[160] I. Rodnianski, W. Schlag, A. Soffer, Dispersive analysis of charge transfer models, *Comm. Pure Appl. Math.* **58** (2005), 149–216.

[161] A. Soffer, Soliton dynamics and scattering, pp. 459–471 in: International Congress of Mathematicians, vol. III, European Mathematical Society, Zürich, 2006.

[162] A. Soffer, M. I. Weinstein, Multichannel nonlinear scattering for nonintegrable equations, *Comm. Math. Phys.* **133** (1990), 119–146.

[163] A. Soffer, M. I. Weinstein, Multichannel nonlinear scattering for nonintegrable equations. II. The case of anisotropic potentials and data, *J. Diff. Equations* **98** (1992), 376–390.

[164] A. Soffer, M. I. Weinstein, Resonances, radiation damping and instability in Hamiltonian nonlinear wave equations, *Invent. Math.* **136** (1999), 9–74.

[165] A. Soffer, M. I. Weinstein, Selection of the ground state for nonlinear Schrödinger equations, *Rev. Math. Phys.* **16** (2004), 977–1071.

[166] D. M. A. Stuart, Modulational approach to stability of non-topological solitons, *J. Math. Pure. Appl.* **80**, no. 1 (2001), 51–83.

[167] T. P. Tsai, Asymptotic dynamics of nonlinear Schrödinger equations with many bound states, *J. Diff. Equations* **192** (2003), 225–282.

[168] T. P. Tsai, H. T. Yau, Classification of asymptotic profiles for nonlinear Schrödinger equations with small initial data, *Adv. Theor. Math. Phys.* **6** (2002), 107–139.

[169] T. P. Tsai, H. T. Yau, Asymptotic dynamics of nonlinear Schrödinger equations: Resonance-dominated and dispersion-dominated solutions, *Comm. Pure Appl. Math.* **55** (2002), 153–216.

[170] M. I. Weinstein, Modulational stability of ground states of nonlinear Schrödinger equations, *SIAM J. Math. Anal.* **16** (1985), 472–491.

Dispersive Decay

[171] S. Agmon, Spectral properties of Schrödinger operators and scattering theory, *Ann. Scuola Norm. Sup. Pisa Cl. Sci. (4)* **2** (1975), 151–218.

[172] P. D'Ancona, L. Fanelli, L. Vega, N. Visciglia, Endpoint Strichartz estimates for the magnetic Schrödinger equation, *J. Funct. Anal.* **258** (2010), 3227–3240.

[173] P. D'Ancona, Kato smoothing and Strichartz estimates for wave equations with magnetic potentials, *Comm. Math. Phys.* **335** (2015), 1–16.

[174] M. Beals, W. Strauss, L^p estimates for the wave equation with a potential, *Comm. Partial Diff. Equations* **18** (1993), 1365–1397.

[175] M. Beceanu, M. Goldberg, Schrödinger dispersive estimates for a scaling-critical class of potentials, *Comm. Math. Phys.* **314** (2012), 471–481.

[176] M. Beceanu, M. Goldberg, Strichartz estimates and maximal operators for the wave equation in $\mathbb{R}^3$, *J. Funct. Anal.* **266** (2014), 1476–1510.

[177] I. Egorova, E. Kopylova, V. A. Marchenko, G. Teschl, Dispersion estimates for one-dimensional Schrödinger and Klein–Gordon equations. Revisited. *Russ. Math. Surv.* **71** (2016), 391–415.

[178] I. Egorova, E. Kopylova, G. Teschl, Dispersion estimates for one-dimensional discrete Schrödinger and wave equations, *J. Spectr. Theory* **5** (2015), 663–696.

[179] M. B. Erdoğan, M. Goldberg, W. R. Green, Dispersive estimates for four-dimensional Schrödinger and wave equations with obstructions at zero energy, *Comm. Partial Diff. Equations* **39** (2014), 1936–1964.

[180] M. Goldberg, W. Schlag, Dispersive estimates for Schrödinger operators in dimensions one and three, *Comm. Math. Phys.* **251** (2004), 157–178.

[181] M. Goldberg, W. R. Green, Dispersive estimates for higher dimensional Schrödinger operators with threshold eigenvalues I: The odd dimensional case, *J. Funct. Anal.* **269** (2015), 633–682.

[182] M. Goldberg, W. R. Green, Dispersive estimates for higher dimensional Schrödinger operators with threshold eigenvalues II: The even dimensional case, *J. Spectr. Theory* **7**, no. 1 (2017), 33–86.

[183] A. Jensen, T. Kato, Spectral properties of Schrödinger operators and time-decay of the wave functions, *Duke Math. J.* **46** (1979), 583–611.

[184] J.-L. Journé, A. Soffer, C. D. Sogge, Decay estimates for Schrödinger operators, *Comm. Pure Appl. Math.* **44** (1991), 573–604.

[185] A. I. Komech, E. Kopylova, Long time decay for 2D Klein–Gordon equation, *J. Funct. Anal.* **259** (2010), 477–502.

[186] A. I. Komech, E. Kopylova, Weighted energy decay for 3D Klein–Gordon equation, *J. Diff. Equations* **248** (2010), 501–520.

[187] A. I. Komech, E. Kopylova, Weighted energy decay for 1D Klein–Gordon equation, *Comm. PDE* **35** (2010), 353–374.

[188] A. I. Komech, E. Kopylova, Dispersion Decay and Scattering Theory, John Wiley, Hoboken, NJ, 2012.

[189] A. I. Komech, E. Kopylova, Dispersion decay for the magnetic Schrödinger equation, *J. Funct. Anal.* **264** (2013), 735–751.

[190] A. I. Komech, E. Kopylova, Weighted energy decay for magnetic Klein–Gordon equation, *Appl. Anal.* **94** (2015), 219–233.

[191] A. Zygmund, Trigonometric Series I, Cambridge University Press, Cambridge, 1968.

[192] A. I. Komech, E. A. Kopylova, M. Kunze, Dispersive estimates for 1D discrete Schrödinger and Klein–Gordon equations, *Appl. Anal.* **85** (2006), 1487–1508.

[193] A. I. Komech, E. A. Kopylova, B. R. Vainberg, On dispersive properties of discrete 2D Schrödinger and Klein–Gordon equations, *J. Funct. Anal.* **254** (2008), 2227–2254.

[194] E. A. Kopylova, Dispersive estimates for discrete Schrödinger and Klein–Gordon equations, *Algebra Anal.* **21** (2009), 87–113.

[195] E. Kopylova, On dispersion decay for 3D Klein–Gordon equation, *Discrete Continuous Dyn. Syst. A* **38** (2018), 5765–5780.

[196] E. A. Kopylova, Dispersion estimates for the Schrödinger and Klein–Gordon equations, *Uspekhi Mat. Nauk* **65** (2010), 97–144.

[197] E. Kopylova, G. Teschl, Dispersion estimates for one-dimensional discrete Dirac equations, *J. Math. Anal. Appl.* **434** (2016), 191–208.

[198] B. Marshall, W. Strauss, S. Wainger, $L^p - L^q$ estimates for the Klein–Gordon equation, *J. Math. Pure. Appl. (9)* **59** (1980), 417–440.

[199] I. Rodnianski, W. Schlag, Time decay for solutions of Schrödinger equations with rough and time-dependent potentials, *Invent. Math.* **155** (2004), 451–513.

[200] E. M. Stein, Harmonic Analysis: Real-Variable Methods, Orthogonality, and Oscillatory Integrals, Princeton Math. Series 43, Princeton University Press, Princeton, NJ, 1993.

[201] D. Tataru, Local decay of waves on asymptotically flat stationary space-times, *Am. J. Math.* **135** (2013), 361–401.

[202] K. Yajima, Dispersive estimates for Schrödinger equations with threshold resonance and eigenvalue, *Comm. Math. Phys.* **259** (2005), 475–509.

Quantum Theory and Classical Electrodynamics

[203] M. Abraham, Prinzipien der Dynamik des Elektrons, *Phys. Z.* **4** (1902), 57–63.

[204] M. Abraham, Theorie der Elektrizität, Bd.2: Elektromagnetische Theorie der Strahlung, Teubner, Leipzig, 1905.

[205] N. Bohr, Discussion with Einstein on epistemological problems in atomic physics, pp. 201–241 in: Schilpp, P. A., ed., Albert Einstein: Philosopher-Scientist, Library of Living Philosophers, Evanston, Illinois, 1949.

[206] A. Einstein, Ist die Trägheit eines Körpers von seinem Energieinhalt abhängig?, *Ann. Phys.* **18** (1905), 639–643.

[207] R. P. Feynman, R. B. Leighton, M. Sands, The Feynman Lectures on Physics, vol. 2, Mainly Electromagnetism and Matter, Addison-Wesley, Reading, MA, 1964.

[208] W. Heisenberg, Der derzeitige Stand der nichtlinearen Spinortheorie der Elementarteilchen, *Acta Phys. Austriaca* **14** (1961), 328–339.

[209] W. Heisenberg, Introduction to the Unified Field Theory of Elementary Particles, Interscience, London, 1966.

[210] L. Houllevigue, L'Évolution des Sciences, A. Collin, Paris, 1908.

[211] C. Itzykson, J. B. Zuber, Quantum Field Theory, McGraw-Hill, New York, 1980.

[212] J. D. Jackson, Classical Electrodynamics, 3rd ed., John Wiley, New York, 1999.

[213] A. I. Komech, Quantum Mechanics: Genesis and Achievements, Springer, Dordrecht, 2013.

[214] A. I. Komech, Quantum jumps and attractors of Maxwell–Schrödinger equations, arXiv 1907.04297 math-ph. https://arxiv.org/abs/1907.04297

[215] J. von Neumann, Mathematical Foundations of Quantum Mechanics, Princeton University Press, Princeton, NJ, 1955.

[216] R. Newton, Quantum Physics, Springer, New York, 2002.

[217] J. J. Sakurai, Advanced Quantum Mechanics, Addison-Wesley, Reading, MA, 1967.

[218] L. I. Schiff, Quantum Mechanics, McGraw-Hill, New York, 1955.

[219] E. Schrödinger, Quantisierung als Eigenwertproblem, *Ann. Phys.* I, II **79** (1926) 361, 489; III **80** (1926) 437; IV **81** (1926) 109. (English translation in E. Schrödinger, Collected Papers on Wave Mechanics, Blackie & Sohn, London, 1928.)

[220] G. K. Woodgate, Elementary Atomic Structure, Clarendon Press, Oxford, 2002.

Omega-Hyperon

[221] V. E. Barnes et al., Observation of a hyperon with strangeness minus three, *Phys. Rev. Lett.* **12** (1964), 204–206.

[222] M. Gell-Mann, Symmetries of baryons and mesons, *Phys. Rev. (2)* **125** (1962), 1067–1084.

[223] F. Halzen, A. Martin, Quarks and Leptons: An Introductory Course in Modern Particle Physics, John Wiley, New York, 1984.

[224] Y. Ne'eman, Unified interactions in the unitary gauge theory, *Nuclear Phys.* **30** (1962), 347–349.

Index

a priori bound, 18, 21, 46, 53, 55, 69, 75, 127
a priori estimate, 1, 32, 46, 58, 67, 77
Abraham, 58, 67, 169
absolutely continuous distribution, 96, 100
adiabatic asymptotics, 167, 168
adiabatic effective dynamics, 175
adiabatic invariant, 119
Agmon, 4, 178, 182
analytic continuation, 98, 132, 155
analytic function, 35–37, 44, 117, 132, 135, 159
analytic properties, 135
Ascoli–Arzelà theorem, 104
asymptotic completeness, 16, 19
asymptotic stability, 114, 120, 166
asymptotic stability of solitary manifold, 10, 12, 114, 122
asymptotic state, 19
asymptotics, 4–7, 9, 10, 16, 46, 60, 76, 80, 88, 89, 112, 115, 116, 118, 120–125, 130, 154, 168, 171, 174, 183, 195–197
attracting set, 37, 50, 65
attraction, 10, 11, 59, 65, 66, 94
  in the mean, 50, 52, 56
  to solitons, 95, 121
  to stationary orbits, 91, 94, 95
  to stationary states, 19, 50
attractor, ix, x, 1–3, 9, 14, 76, 192, 193

Bambusi, 79
Banach phase space, 2, 111
Banach space, ix, 6, 10, 127, 136
baryon, 10, 112, 113
Berestycki, 8

Berezin, 68
Bethe, 68
Bohr, 192, 198
Bohr's postulates, 11, 192–194, 197, 198
Bohr's quantum jumps, ix, 1, 5, 11, 25, 192–195
Bohr's rule, 198
Born, 193
Born probabilistic interpretation, 198
Born series, 184
bound component, 102, 103
boundary value problem, 45, 48
Buslaev, 4, 10, 116, 124

canonical form, 88, 167
canonical transformation, 77, 79, 80, 88
Cauchy data, 40, 46, 53, 55
Cauchy problem, 16–18, 21, 30, 43, 53, 70, 72, 73
Cauchy residue theorem, 101, 132, 134
Cauchy–Schwarz inequality, 22, 32, 40, 47, 63, 102
Cazenave, 3
charge conservation, 126, 138
charge density, 13, 56, 58, 67, 77, 169, 194
chemical reaction, 2
Cherenkov radiation, 122
Chern–Simon–Schrödinger system, 168
chromodynamics, 193
classical electrodynamics, 1, 5, 13, 59, 67, 77, 169
classical particle, 175
Comech, 1, 4
comoving frame, 6, 7, 77–79, 121
complexification, 117

212

concentrated nonlinearity, 8, 13, 68, 91, 94
concentration compactness, 123
confining potential, 44, 57, 67, 68
conjecture on attractors, ix, 4, 10, 111–113, 169, 192, 195
continuous spectrum, 9, 59, 96, 99, 102, 108–110, 115, 118, 121, 122, 125, 134, 141, 147, 165, 173, 181, 197, 198
convex hull, 106
convolution, 59, 64, 104, 107, 109
convolution representation, 63, 67
Coulomb potential, 9, 58, 196
critical energy, 123
critical focusing nonlinear Schrödinger equation, 123
critical point, 9, 79, 81, 82
current density, 67, 194, 197

d'Alembert, 17, 27
    equation, 6, 13, 26
    formula, 14, 20, 40, 41, 47
    operator, 194
    representation, 20, 30, 32, 40, 41, 52
de Broglie, 193
defocusing nonlinearity, 3
determinant, 159
difference scheme, 8, 91, 94, 174
diffraction theory, 98
Dirac, 68, 195
    distribution, 134, 136, 139
    equation, 8, 91, 94, 111, 121, 178
    matrices, 94
discrete equation, 178
discrete set, 19, 22, 25, 31, 34, 36, 37, 44, 46, 48, 59, 66, 76, 176
discrete spectrum, 8, 9, 91, 114, 118, 129, 133, 136, 147, 161, 162, 173, 198
dispersion relation, 173
dispersive component, 102
dispersive decay, ix, 11, 90, 178–181
dispersive mechanism, 108
dispersive radiation, 108, 111
dispersive wave, 7, 9, 16, 73, 89, 90, 102, 109, 123, 171, 173–177, 196
dissipative system, 1, 2, 76
distribution, x, 14, 17, 18, 30, 97, 102, 104, 107, 124, 129, 132, 136, 182, 191
Duhamel representation, 47, 90, 139, 149
dynamical group, 16, 19, 47, 54, 118, 124, 129, 132, 138, 139, 152, 153
dynamical interpretation, 194

dynamical system, 43
dynamical treatment of Bohr's postulates, 192
dynamics, 10, 16, 31, 46, 115, 125, 194
Dynkin scheme, 113

effective dispersion relation, 168
effective dynamics, ix, 11, 166–168, 170, 174–176
effective Hamiltonian equation, 167
effective Hamiltonian functional, 166, 167
effective potential, 116
eigenfunction, 4, 9, 59, 179, 193, 197
eigenfunction expansion, 119, 122
eigenvalue, 8, 10, 114, 118, 122, 129, 171, 173, 178–182
eightfold way, 113
Einstein, 168, 169, 198
Einstein equations, 123
Einstein–Dirac system, 168
elementary particle, 111, 113, 198
embedded eigenvalues, 8, 91, 96, 102
energy, 46, 47, 57, 62, 72, 87, 88, 123, 168, 169, 179
    absorption, 76
    conservation, 2, 3, 5, 6, 18, 21, 31, 32, 44, 46, 58, 69, 70, 74, 76, 79, 83, 92, 116, 138
    density, 177
    dissipation, 2
    flow, 42
    functional, 42, 52
    norm, 16
    radiation, 2–6, 13, 21, 25, 41, 42, 45, 52, 59, 62, 63, 67, 76, 108–110, 118, 176
    transfer, 108, 109
entire function, 106
equations with delay, 2, 28
equations with memory, 2
Esteban, 8, 196
example, 4, 6, 13, 19, 23, 27, 29, 35–37, 46, 60, 112, 119, 121–123, 137, 138, 173, 180, 183, 188, 197
exceptional equation, 112
extended electron, 58, 67, 169
external potential, 57, 166, 194–196

Faddeev, 68
Fermi, 68

Fermi Golden Rule, 59, 118, 120, 122
finite-energy solution, 3, 4, 9, 16, 26, 28, 58, 66, 67, 116, 123–126, 170, 171, 194, 195, 197
finite-energy state, 16, 29, 67, 87, 118, 124, 173
form-factor, 59
Fourier integral, 100
Fourier representation, 101, 129
Fourier transform, x, 58, 78, 96–99, 104, 106, 109, 154, 180
Fourier–Laplace transform, 97, 102, 108
free Klein–Gordon equation, 179
free Schrödinger equation, 118, 124, 129, 152, 196
free Schrödinger operator, 182
free wave equation, 16, 19, 61, 69, 74, 87, 89, 90
friction, 1, 2, 24, 25
fundamental solution, 109

G-invariance, 111
G-invariant equation, 8, 111, 112, 195
Galgani, 79
Galilean transformation, 117
Gell-Mann, 10, 112
generator, 10, 100, 109, 112, 113, 118, 119, 125
generic equations, 4, 9, 11, 76, 111, 112
Georgiev, 8, 9, 196
Ginzburg–Landau equation, 12, 120, 173
Ginzburg–Landau potential, 23, 93, 122, 170, 173
global attraction, 2, 3, 10, 11
    to N-frequency trajectories, 9
    to solitary manifold, 7
    to solitons, ix, 6, 7, 11, 77, 90, 170
    to stationary orbits, ix, 8, 91, 193, 194
    to "stationary SO(3)-orbits", 9
    to stationary states, ix, 2, 4, 6, 13–15, 19, 28, 37, 43, 45, 56, 66–68, 76
global attractor, ix, 1–3, 7, 15, 18, 24, 31, 45, 76, 112, 192, 194, 199
global energy norm, 16, 19, 89, 196
global minimum, 79, 81
global norm, 5, 6, 76, 118, 124
Green function, 68
Grillakis, 9
Gronwall inequality, 55
ground state, 114, 115, 123
group, 38, 128, 133–135, 149, 179

group of translations, 112
group velocity, 173, 176

Hamiltonian dynamics, 166, 175
Hamiltonian equation, 167
Hamiltonian functional, 6, 14, 17, 30, 43, 57, 58, 67, 77, 79, 81, 83, 87, 88, 92, 114, 126, 128, 129, 167, 170, 174
Hamiltonian operator, 4, 119, 122
Hamiltonian PDEs, ix, x, 1–3
Hamiltonian structure, 88, 111, 167
Hamiltonian system, 3, 14, 15, 17, 24, 30, 67, 76, 77, 81, 87, 88, 92, 124, 126, 128, 167, 170, 174
Haraux, 3
harmonic, 110
harmonic analysis, x, 4, 106
harmonic source, 108, 109
Heisenberg, 193, 198
Helmholtz equation, 98
Hessian, 129
higher harmonics, 108, 111
Hilbert manifold, 69
Hilbert phase space, 2, 16, 17, 29, 30, 44, 57, 67, 78, 80, 81, 92, 111, 117–119, 124, 128
Hilbert space, ix, 6, 14, 16, 29, 54, 57, 102
Hopf, 3
hyperbolic PDEs, 3

implicit function theorem, 143
incident wave, 20, 41
inflation of spectrum, 109–111
initial data, 3, 6, 14, 42, 46, 53, 69, 71–74, 77, 79, 84, 86, 87, 115, 116, 118, 121, 142–144, 154, 173
integral kernel, 154

Jörgens, 3
Jacobian matrix, 143
Jensen, 4, 125, 178, 181, 182
Journé, 178

Kato, 4, 125, 181, 182
Kato theorem, 96
kernel, 154
Kerr black hole, 123
Kerr solutions, 123
kink, 123, 171–174
Kirchhoff, 61
Kirchhoff formula, 84

Klein–Gordon equation, 8, 10, 90, 91, 94, 96, 108, 109, 111, 121, 173, 178–180
Klein–Gordon–Dirac system, 9
Klein–Gordon–Maxwell system, 168
Kopylova, 1, 4
Krein theory, 4
Krein–Langer theory, 119, 122

Lagrangian, 87
Lagrangian functional, 87
Lamb, 15
Landau, 1
Laplace representation, 135
Laplace transform, 131, 132
Larmor formula, 67
laser radiation, 193
Lax, 3
Legendre transformation, 87, 88
Liénard formula, 67
Liénard–Wiechert asymptotics, 60, 67
Liénard–Wiechert formulas, 59
Lie algebra, 4, 113
Lie group, 112
Lie symmetry group, 4, 111, 113
limiting absorption principle, 96, 98
limiting amplitude, 4, 109
limiting amplitude principle, 108
linear eigenvalue problem, 196
linear hyperbolic equation, 3
linear Schrödinger equation, 9, 198
linearization, 117, 122, 125, 130, 133, 173
linearized dynamics, 10, 115, 119, 122, 124
linearized equation, 11, 117, 122, 125, 134, 138, 139, 144, 154, 171, 173
linearized operator, 129
Lions J.-L., 3
Lions P.-L., 8
local attraction, 10, 114
local energy decay, 3
local energy norm, 195
local energy seminorm, 17, 109
local seminorm, 5, 7, 76
long-time asymptotics, 3, 4, 9, 11, 37, 50, 109, 113, 123, 198
long-time attraction, 115
long-time convergence, 122
long-time decay, 88
Lorentz contraction, 171
Lorentz transformation, 171
lower harmonics, 108
Lusternik–Schnirelmann theory, 9

Lyapunov, 10, 11
Lyapunov function, 120

majorant, 120, 124, 138, 151
mass–energy equivalence, 11, 166, 168, 169
Maxwell equations, 10, 67, 123, 194, 196
Maxwell field, 77
Maxwell potentials, 194
Maxwell–Dirac system, 8, 192, 195, 196, 199
Maxwell–Lorentz equations, 66, 79, 88, 90, 168, 169
Maxwell–Lorentz equations with rotating charge, 9, 168
Maxwell–Schrödinger system, 9, 12, 192–194, 196–199
measure, 104
metastable tori, 122
method of compactness, 1, 3
metric, 17, 18, 51, 57, 59
modulation equations, 11, 115, 119, 125, 140–142
momentum, 80–83, 88, 166, 167
momentum conservation, 77, 79–81, 83
Morawetz, 3
multiphoton radiation, 121
multiplier, 180
multiplier of quasimeasures, x, 4, 8, 91, 99, 105

Navier–Stokes equations, 1, 2
Ne'eman, 10, 112
Noja, 68
nondiscrete set, 35
nonlinear dynamics, 110
nonlinear eigenfunction, 124
nonlinear eigenvalue problem, 4, 93, 112, 126, 193, 195, 196
nonlinear Goursat problem, 53–55
nonlinear Hamiltonian equation, 3, 4, 11, 13, 14, 192
nonlinear Hartree equation, 168
nonlinear Helmholtz equation, 106
nonlinear Kato theorem, 8, 91, 99, 102
nonlinear Klein–Gordon equation, 90, 91, 115, 122
nonlinear Lamb system, 14, 15, 28
nonlinear oscillator, 8, 10, 14, 25, 28, 29, 68, 91, 94, 114, 121, 124
nonlinear radiative mechanism, 91, 107, 111
nonlinear scattering, 15

nonlinear Schrödinger equation, 94, 115, 122, 124, 168
nonlinear wave equation, 13, 28, 43, 53, 79, 115, 122, 123, 168, 170, 174
nonlinearity, 19, 43, 70, 117, 125, 173
nonlocal interaction, 94
nonresonance conditions, 122
Nöther theorem, 126
nuclear reactions, 199
nuclei, 199
nuclei classification, 199
numerical experiment, 111
numerical simulation, ix, 11, 168, 170, 171, 173, 175, 193

omega-compact trajectory, 95, 96
omega-compactness, 95, 96, 103
omega-limit point, 45
omega-limit trajectory, 8, 13, 45, 91, 95, 96, 105, 109, 110
omega-set, 38
orbit, 7, 23, 38, 50, 128, 192
orbital stability, 9, 79, 80, 83, 84, 120, 125, 129
orthogonal symmetry group, 4, 9
orthogonality condition, 115, 141, 142, 179
oscillator, 15, 20, 23, 25, 26, 124–126
oscillatory integral, 135

Paley–Wiener estimates, 100
Parseval–Plancherel identity, 82, 100, 104, 153
particle, 15, 56, 58, 59, 67, 113, 121, 169
particle momentum, 167
Peierls, 68
Perelman, 10, 116, 124
perturbation, 59, 123
perturbation theory, 118, 121, 141, 197–199
Phillips, 3
Poincaré, 120
Poincaré normal form, 120
point nonlinearity, 14
point particle, 67
polynomial, 35, 127
polynomial nonlinearity, 170, 173
polynomial potential, 174
Posilicano, 68
potential, 9, 23, 32, 35–37, 46, 57, 67, 68, 70, 77, 84–87, 114–117, 122, 125, 126, 170, 172, 173, 175, 177–181, 194, 197

potential energy, 32, 44
probabilistic interpretation, 193
projection operator, 141

quantum
   amplifier, 193
   chromodynamics, 113
   field theory, 193, 199
   jumps, 194
   mechanics, 192
   postulate, 192, 193, 198
   stationary orbit, 193
   stationary state, ix, 124, 192, 193, 199
   theory, ix
quantum postulate, 2
quantum stationary state, 11
quantum theory, 2, 193, 198, 199
quarks, 113
quasi-periodic function, 10
quasimeasure, 96, 97, 99, 103, 104

radiation, 26, 109, 175–177, 192, 198
radiation damping, 1, 5, 13, 59, 66, 67, 77
radiation integral, 62
radiation power, 67
radiative mechanism, 173
real-analytic function, 31, 34, 44, 49
reduced equation, 20, 22, 23, 25
reduction of spectrum, 96, 104, 105
reflected wave, 20
relative equilibria, 120
relativistic equation, 7, 8, 11, 12, 90, 95, 111, 120, 122, 168, 170, 174, 196
relativistic kinetic energy, 56
relativistic particle, 7, 10, 13, 56, 66, 77
Relativity General Theory, 123
Relativity Special Theory, 169
relaxation, 2, 22, 28, 39, 40, 42, 43
relaxation of acceleration, 59, 60, 63, 64, 67, 79, 80, 84, 86
resolvent, 109, 125, 132, 134, 135, 154, 158, 159, 182, 183
resonance, 114, 118, 122, 178–181
retarded potential, 59, 60, 67, 86
Reynolds number, 1
Riemann–Lebesgue theorem, 102, 105, 154
Riesz interpolation theorem, 179, 180
Riesz projection, 132, 181
rotation group, 124, 126
rotation symmetry, 126

scalar wave field, 77

scattering, 3, 4, 16, 123
  asymptotics, 19
  operator, 19
  state, 16, 118, 124, 130
  theory, 3

Schrödinger, 11, 194
  equation, 11, 107, 112, 114, 117, 121, 122, 124, 130, 139, 178–181, 194, 195, 197
  operator, 109, 114, 115, 173, 181, 182, 193, 197
  theory, 193, 194, 197

Schwarzschild black hole, 123

Segal, 3

self-energy, 58, 169

seminorm, 29, 44, 52, 57, 59

separatrix, 2

Seré, 8, 196

Shatah, 9

Sigal, 118

single-frequency asymptotics, 7, 195

single-frequency spectrum, 109

slowly varying external potential, 166, 168, 174

Sobolev embedding theorem, 1, 30, 55, 74

Sobolev space, x, 39, 97, 127, 180

Soffer, 10, 11, 90, 114, 178

Sogge, 11, 178

solitary manifold, 8, 10, 11, 78, 93, 94, 115, 118, 119, 121, 124, 125, 133, 166, 167, 175

solitary wave, 126

soliton, ix, 10, 11, 78–84, 86, 90, 114, 117, 119–123, 125, 140, 142, 152, 166–171, 174–177, 196

soliton asymptotics, 84, 152, 174, 196

soliton-like asymptotics, 166

soliton-like solutions, 174, 175

spectral density, 96

spectral Fourier representation, 182

spectral gap, 93, 96, 104, 107

spectral gap infinite, 95

spectral inclusion, 105–108

spectral representation, 9, 96, 99, 125, 184

spectrum, 10, 96, 118, 165, 173

spectrum absolute continuous, 120

Spohn, 1, 4

spreading of spectrum, 108

stationary equation, 44, 48, 78, 116

stationary G-orbit, 8, 9, 112, 195, 196

stationary orbit, 7–9, 93, 95, 96, 114–117, 122, 124–130, 133, 142, 192–194

stationary solution, 6, 34, 36, 48, 68, 81, 123, 171, 173

stationary state, 2, 5, 16, 18, 19, 25, 31, 33, 44, 45, 48, 58, 59, 74, 76, 170

stochastic Lamb system, 15

Strauss, 3, 8, 9

Strichartz estimate, 179

strictly nonlinear equation, 8, 9, 91, 93

string, 14, 15, 23, 28, 29

strong Huygens principle, 6, 74, 79, 80, 84, 87, 88

strong interaction, 10, 113

Stuart, 1, 4, 10

Sulem, 10, 116

symmetry, 46, 129

symmetry group, 4, 9, 10, 111–113, 117, 128, 195

symmetry group of translations, 4, 6

symplectic form, 119, 128, 129, 134, 141

symplectic manifold, 119

symplectic projection, 116, 118, 120, 133, 134, 147

symplectic structure, 80, 82, 87

tangent space, 118, 119, 129, 133, 134

tempered distribution, 96, 97, 99, 100

threshold frequency, 109

time delation, 171, 173

Titchmarsh, 106

Titchmarsh convolution theorem, x, 1, 4, 8, 91, 94, 96, 106–108, 111

total momentum, 167

transition, 5, 25, 192

translation-invariant equation, 116, 118, 120, 196

transversal component, 11, 119, 120, 125, 142, 143

transversal dynamics, 142

transversal subspace, 119

trivial group, 112

trivial symmetry group, 4, 76

truncated resolvent, 181

Tsai, 10

U(1)-invariant oscillator, 121

ultraviolet divergence, 58

unitary group, 112
unitary symmetry group, 4, 7

Vainberg, 1, 3, 4
Van der Corput lemma, 186, 191
variation, 33, 48
variation equation, 33, 48
Vinnichenko, 170
von Neumann, 197

wave equation, 2, 3, 6, 10, 13, 15, 46, 56, 58, 68, 78–80, 86, 91, 122, 123, 178, 179
wave field, 108, 109
wave mapping, 123
wave packet, 173, 176
wave–particle duality, 193, 198
wave–particle system, 56, 77
waveguide, 109
weak compactness, 102

weak convergence, 102
weak coupling, 89
weighted norm, 115, 121
weighted Sobolev norm, 178
Weinstein, 10, 114
well-posedness, 3, 20, 32, 45, 46, 69, 70, 72, 123
Wiener condition, 13, 59, 60, 64, 67, 76–78, 94, 118, 121
Wiener Tauberian theorem, x, 1, 4, 13, 59, 64, 67, 77
Wigner, 68

Yafaev, 68
Yau, 10

Zeidler, 68
Zeldovich, 68
Zygmund lemma, 125, 135, 154

Printed in the United States
by Baker & Taylor Publisher Services